# Ultrafast Ultrasound Imaging

# Ultrafast Ultrasound Imaging

Special Issue Editors

**Hideyuki Hasegawa**
**Chris L. de Korte**

MDPI • Basel • Beijing • Wuhan • Barcelona • Belgrade

**MDPI**

*Special Issue Editors*

Hideyuki Hasegawa
University of Toyama
Japan

Chris L. de Korte
Radboud University Medical Centre
The Netherlands

*Editorial Office*
MDPI
St. Alban-Anlage 66
Basel, Switzerland

This is a reprint of articles from the Special Issue published online in the open access journal *Applied Sciences* (ISSN 2076-3417) from 2017 to 2018 (available at: http://www.mdpi.com/journal/applsci/special_issues/Ultrafast_Ultrasound_Imaging)

For citation purposes, cite each article independently as indicated on the article page online and as indicated below:

LastName, A.A.; LastName, B.B.; LastName, C.C. Article Title. *Journal Name* **Year**, *Article Number, Page Range.*

**ISBN 978-3-03897-127-6 (Pbk)**
**ISBN 978-3-03897-128-3 (PDF)**

# Contents

# About the Special Issue Editors

**Hideyuki Hasegawa**, Ph.D., received his B.E. degree in 1996 and his Ph.D. degree in 2001 both from Tohoku University, Sendai, Japan in 2001. He became an assistant professor in 2002 and associate professor in 2007 at the Graduate School of Engineering, Tohoku University. Since 2015, he is a professor at the Graduate School of Science and Engineering, the University of Toyama. His research interests include medical ultrasound and its application to functional imaging. Prof. Hasegawa is a member of the IEEE, the Acoustical Society of Japan, the Japan Society of Ultrasonics in Medicine, and the Institute of Electronics, Information and Communication Engineers.

**Chris L. de Korte**, Prof.Dr., is Full Professor in Medical Ultrasound Techniques and Chair of the Medical UltraSound Imaging Centre at the Department of Radiology and Nuclear Medicine of Radboud University Medical Center. His research is on functional imaging using ultrasound with a focus on cardiovascular applications. He studied Electrical Engineering at the Eindhoven University of Technology and, in 1999, obtained his Ph.D. at the Biomedical Engineering Group of the Thoraxcentre, Erasmus University Rotterdam with his thesis on Intravascular Ultrasound Elastography. In 2002, he joined the Clinical Physics Laboratory, Department of Pediatrics of the Radboud University, Nijmegen Medical Center of which he became the Head in 2004. In 2006, he was registered as a Medical Physicist. Prof. de Korte is President of the Netherlands Society for Medical Ultrasound (NVMU) and the national delegate of the European Federation of Societies for Ultrasound in Medicine and Biology (EFSUMB). He is an associate editor for IEEE Transactions UFFC and the Journal of Medical Ultrasonics. He serves as an editorial board member of Ultrasound In Medicine and Biology and the Journal of the British Medical Ultrasound Society and the Technical Program Committee of the International IEEE Ultrasonics Conferences.

*applied*
*sciences*

MDPI

*Editorial*

# Special Issue on Ultrafast Ultrasound Imaging and Its Applications

**Hideyuki Hasegawa [1],* and Chris L. de Korte [2],***

[1]   Graduate School of Science and Engineering, University of Toyama, Toyama 930-8555, Japan
[2]   Medical UltraSound Imaging Center (MUSIC 766), Radboudumc, P.O. Box 1738, 6500HB Nijmegen, The Netherlands
*    Correspondence: hasegawa@eng.u-toyama.ac.jp (H.H.); Chris.deKorte@radboudumc.nl (C.L.d.K.)

Received: 4 July 2018; Accepted: 6 July 2018; Published: 10 July 2018

## 1. Introduction

Among medical imaging modalities, such as computed tomography (CT) and magnetic resonance imaging (MRI), ultrasound imaging stands out in terms of temporal resolution. Due to the nature of medical ultrasound imaging, it has been used for observation of the morphology of living organs and, also, functional imaging, such as blood flow imaging and evaluation of the cardiac function. Ultrafast ultrasound imaging, which has become practically available recently, significantly increases the possibilities of medical ultrasounds for functional imaging. Ultrafast ultrasound imaging realizes typical imaging frame-rates up to ten thousand frames per second (fps). Owing to such an extremely high temporal resolution, ultrafast ultrasound imaging enables visualization of rapid dynamic responses of biological tissues which cannot be observed and analyzed by conventional ultrasound imaging. Various studies have been conducted to make ultrafast ultrasound imaging useful in clinical practice and, also, for further improvements in the performance of ultrafast ultrasound imaging itself as well as finding new potential applications.

## 2. Ultrafast Ultrasound Imaging

The primary factor limiting the temporal resolution in ultrasound imaging is the speed of sound in the body. Ultrasound pulses can be transmitted at pulse repetition frequencies of about 10,000 Hz for superficial organs and about 5000 Hz for deep organs. An ultrasound image is typically composed of 100–250 scan lines and one transmit-receive event is required to create one scan line because a focused transmit beam is used in conventional ultrasound imaging (ultrasonic echoes are coming from only a limited region). As a result, the imaging frame rate is limited to less than 100 fps unless the number or density of scan lines is not reduced.

The concept of ultrafast ultrasound imaging is not new. It was first developed and examined in the 1970s [1–3]. In ultrafast ultrasound imaging, unfocused transmit beams, such as plane and diverging beams, are used. As a result, ultrasonic echoes are coming from a wide region illuminated by an unfocused transmit beam. By creating focused beams in reception, a number of scan lines can be created simultaneously. Therefore, the number of emissions required to create one image frame can be reduced significantly. In an extreme case, an ultrasound image can be created by only one transmit-receive event if we can illuminate a region of interest by a single emission of an unfocused transmit beam.

On the other hand, image quality in ultrafast ultrasound imaging, e.g., lateral spatial resolution and contrast, is inherently worse than that in conventional imaging using focused transmit beams because the directivity is created only in reception and ultrasonic echoes from a wide region produce undesirable echoes. Various attempts have been made for improvement in image quality in ultrafast ultrasound imaging. Spatial coherent compounding is a frequently used method to improve the image quality in ultrafast ultrasound imaging [4–6]. By compounding point spread functions (PSFs) created from multiple steered beams, the compounded PSF is sharpened because only the central parts of the PSFs are coherently summed, and other parts of the PSFs are incoherently summed (canceled). The improvement of the image quality is increased by increasing the number of coherently compounded angles and consequently the imaging frame rate is degraded.

Another approach is to improve the performance of an ultrasound beamformer. One strategy is to use the coherence among ultrasonic echo signals received by individual transducer elements [7,8]. Such methods utilize the characteristics of received signals, e.g., echoes from a focal point are temporally aligned after delay compensation done by a delay-and-sum (DAS) beamformer, while out-of-focus echoes are not aligned. Coherence evaluation metrics, such as coherence factor (CF) and phase coherence factor (PCF), were developed and demonstrated to improve ultrasound image quality. Adaptive beamforming is an alternative strategy for improvement in the performance of an ultrasonic beamformer. The minimum variance beamformer [9] was introduced in medical ultrasound imaging in the late 2000s [10,11]. It minimizes the power of received ultrasonic signals (undesired echoes and noise are suppressed) while keeping the all-pass characteristic with respect to the desired direction (focal point). Significant improvements in image quality can be realized by minimum variance beamforming. On the other hand, the computational complexity of the minimum variance beamformer is very high, and developments of efficient implementations of the minimum variance beamformer are still ongoing [12,13]. In addition, various studies on improvement in the performance of the minimum variance beamformer have been conducted [14,15].

## 3. Applications and Ongoing Developments

As described above, the basic principle of ultrafast ultrasound imaging was developed in the 1970s. However, practical applications of ultrafast ultrasound imaging only arise from the early 2000s. Ultrafast ultrasound imaging was first used for visualizing the propagation of a shear wave induced by acoustic radiation force applied by an ultrasonic push pulse [16]. The measurement of shear wave propagation speed is useful for evaluation of the elastic properties of biological tissues. Shear wave imaging had a great impact on the field of medical ultrasonics. Owing to the extremely high temporal resolution of ultrafast ultrasound imaging, various applications have been developed for functional imaging, such as blood flow imaging [16–20], evaluation of cardiac function [21–23], and vascular viscoelastic properties [24–26].

Ultrafast ultrasound imaging has not been used practically for very long due to its limited image quality and hardware limitations. However, it attracts significant attention because its extremely high temporal resolution is of great benefit for measurements of tissue dynamic properties. Ultrafast functional ultrasound imaging is now moving to 3D imaging. Various developments are ongoing for transducer fabrication, large-scale acquisition systems, beamforming in 3D space, and estimation of 3D tissue functional properties.

**Acknowledgments:** We would like to thank talented authors, professional reviewers, and a dedicated editorial team of Applied Sciences. This issue would not be possible without their contributions. In addition, we would like to give special thanks to Ms. Felicia Zhang from MDPI Branch Office, Beijing.

**Conflicts of Interest:** The authors declare no conflict of interest.

## References

1. Bruneel, C.; Torguet, R.; Rouvaen, K.M.; Bridoux, E.; Nongaillard, B. Ultrafast echotomographic system using optical processing of ultrasonic signals. *Appl. Phys. Lett.* **1977**, *30*, 371–373. [CrossRef]
2. Delannoy, B.; Torguet, R.; Bruneel, C.; Bridoux, E.; Rouaven, J.M.; Lasota, H. Acoustical image reconstruction in parallel-processing analog electronic systems. *J. Appl. Phys.* **1979**, *50*, 3153–3159. [CrossRef]
3. Shattuck, D.P.; Weinshenker, M.D.; Smith, S.W.; von Ramm, O.T. Explososcan: A parallel processing technique for high speed ultrasound imaging with linear phased arrays. *J. Acoust. Soc. Am.* **1984**, *75*, 1273–1282. [CrossRef]

4.  Montaldo, G.; Tanter, M.; Bercoff, J.; Benech, N.; Fink, M. Coherent plane-wave compounding for very high frame rate ultrasonography and transient elastography. *IEEE Trans. Ultrason. Ferroelectr. Freq. Control.* **2009**, *56*, 489–506. [CrossRef] [PubMed]

5.  Couade, M.; Pernot, M.; Tanter, M.; Messas, E.; Bel, A.; Ba, M.; Hagege, A.-A.; Fink, M. Ultrafast imaging of the heart using circular wave synthetic imaging with phased arrays. In Proceedings of the IEEE International Ultrasonics Symposium, Rome, Italy, 20–23 Septembar 2009; pp. 515–518.

6.  Hasegawa, H.; Kanai, H. High-frame-rate echocardiography using diverging transmit beams and parallel receive beamforming. *J. Med. Ultrason.* **2011**, *38*, 129–140. [CrossRef] [PubMed]

7.  Li, P.-C.; Li, M.-L. Adaptive imaging using the generalized coherence factor. *IEEE Trans. Ultrason. Ferroelectr. Freq. Control* **2003**, *50*, 128–141. [PubMed]

8.  Camacho, J.; Parrilla, M.; Fritsch, C. Phase coherence imaging. *IEEE Trans. Ultrason. Ferroelectr. Freq. Control* **2009**, *56*, 958–974. [CrossRef] [PubMed]

9.  Capon, J. High-resolution frequency-wavenumber spectrum analysis. *Proc. IEEE* **1969**, *57*, 1408–1418. [CrossRef]

10. Synnevåg, J.F.; Austeng, A.; Holm, S. Adaptive beamforming applied to medical ultrasound imaging. *IEEE Trans. Ultrason. Ferroelectr. Freq. Control* **2007**, *54*, 1606–1613. [CrossRef] [PubMed]

11. Holfort, I.K.; Gran, F.; Jensen, J.A. Broadband minimum variance beamforming for ultrasound imaging. *IEEE Trans. Ultrason. Ferroelectr. Freq. Control* **2009**, *56*, 314–325. [CrossRef] [PubMed]

12. Synnevåg, J.F.; Austeng, A.; Holm, S. A low-complexity data dependent beamformer. *IEEE Trans. Ultrason. Ferroelectr. Freq. Control* **2011**, *58*, 281–289. [CrossRef] [PubMed]

13. Asl, B.M.; Mahloojifar, A. A low-complexity adaptive beamformer for ultrasound imaging using structured covariance matrix. *IEEE Trans. Ultrason. Ferroelectr. Freq. Control* **2012**, *59*, 660–667. [CrossRef] [PubMed]

14. Asl, B.M.; Mahloojifar, A. Minimum variance beamforming combined with adaptive coherence weighting applied to medical ultrasound imaging. *IEEE Trans. Ultrasonics. Ferroelectr. Freq. Control* **2009**, *56*, 1923–1931. [CrossRef] [PubMed]

15. Kim, K.; Park, S.; Kim, J.; Park, S.-B.; Bae, M. A fast minimum variance beamforming method using principal component analysis. *IEEE Trans. Ultrason. Ferroelectr. Freq. Control* **2014**, *61*, 930–945. [CrossRef] [PubMed]

16. Udesen, J.; Gran, F.; Lindskov Hansen, K.; Jensen, J.A.; Thomsen, C.; Nielsen, M.B. High frame-rate blood vector velocity imaging using plane waves: Simulations and preliminary experiments. *IEEE Trans. Ultrason. Ferroelectr. Freq. Control* **2008**, *55*, 1729–1743. [CrossRef] [PubMed]

17. Hasegawa, H.; Kanai, H. Simultaneous imaging of artery-wall strain and blood flow by high frame rate acquisition of RF signals. *IEEE Trans. Ultrason. Ferroelectr. Freq. Control* **2008**, *55*, 2626–2639. [CrossRef] [PubMed]

18. Bercoff, J.; Montaldo, G.; Loupas, T.; Savery, D.; Mézière, F.; Fink, M.; Tanter, M. Ultrafast compound Doppler imaging: Providing full blood flow characterization. *IEEE Trans. Ultrason. Ferroelectr. Freq. Control* **2011**, *58*, 134–147. [CrossRef] [PubMed]

19. Yiu, B.Y.; Yu, A.C. High-frame-rate ultrasound color-encoded speckle imaging of complex flow dynamics. *Ultrasound Med. Biol.* **2013**, *39*, 1015–1025. [CrossRef] [PubMed]

20. Takahashi, H.; Hasegawa, H.; Kanai, H. Echo speckle imaging of blood particles with high-frame-rate echocardiography. *Jpn. J. Appl. Phys.* **2014**, *53*, 07KF08. [CrossRef]

21. Honjo, Y.; Hasegawa, H.; Kanai, H. Two-dimensional tracking of heart wall for detailed analysis of heart function at high temporal and spatial resolutions. *Jpn. J. Appl. Phys.* **2010**, *49*, 07HF14. [CrossRef]

22. Provost, J.; Nguyen, V.T.-H.; Legrand, D.; Okrasinski, S.; Costet, A.; Gambhir, A.; Garan, H.; Konofagou, E.E. Electromechanical wave imaging for arrhythmias. *Phys. Med. Biol.* **2011**, *56*, L1. [CrossRef] [PubMed]

23. Cikes, M.; Tong, L.; Sutherland, G.R.; D'hooge, J. Ultrafast cardiac ultrasound imaging: Technical principles, applications, and clinical benefits. *JACC Cardiovasc. Imaging* **2014**, *7*, 812–823. [CrossRef] [PubMed]

24. Shahmirzadi, D.; Li, R.X.; Konofagou, E.E. Pulse-wave propagation in straight-geometry vessels for stiffness estimation: Theory, simulations, phantoms and in vitro findings. *J. Biomech. Eng.* **2012**, *134*, 114502. [CrossRef] [PubMed]

25. Hasegawa, H.; Hongo, K.; Kanai, H. Measurement of regional pulse wave velocity using very high frame rate ultrasound. *J. Med. Ultrason.* **2013**, *40*, 91–98. [CrossRef] [PubMed]
26. Saris, A.E.C.M.; Hansen, H.H.G.; Fekkes, S.; Nillesen, M.M.; Rutten, M.C.M.; de Korte, C.L. A comparison between compounding techniques using large beam-steered plane wave imaging for blood vector velocity imaging in a carotid artery model. *IEEE Trans. Ultrason. Ferroelectr. Freq. Control* **2016**, *63*, 1758–1771. [CrossRef] [PubMed]

![applied sciences logo]

*applied sciences*

MDPI

*Article*

# Adaptive Beamformer Combined with Phase Coherence Weighting Applied to Ultrafast Ultrasound

**Michiya Mozumi [1] and Hideyuki Hasegawa [2,*]**

[1] Faculty of Engineering, University of Toyama, Toyama 930-8555, Japan; s1470267@ems.u-toyama.ac.jp
[2] Graduate School of Science and Engineering, University of Toyama, Toyama 930-8555, Japan
* Correspondence: hasegawa@eng.u-toyama.ac.jp; Tel.: +81-76-445-6741

Received: 14 December 2017; Accepted: 25 January 2018; Published: 30 January 2018

**Abstract:** Ultrafast ultrasound imaging is a promising technique for measurement of fast moving objects. In ultrafast ultrasound imaging, the high temporal resolution is realized at the expense of the lateral spatial resolution and image contrast. The lateral resolution and image contrast are important factors determining the quality of a B-mode image, and methods for improvements of the lateral resolution and contrast have been developed. In the present study, we focused on two signal processing techniques; one is an adaptive beamformer, and the other is the phase coherence factor (PCF). By weighting the output of the modified amplitude and phase estimation (mAPES) beamformer by the phase coherence factor, image quality was expected to be improved. In the present study, we investigated how to implement the PCF into the mAPES beamformer. In one of the two examined strategies, the PCF is estimated using element echo signals before application of the weight vector determined by the adaptive beamformer. In the other strategy, the PCF was evaluated from the element signals subjected to the mAPES beamformer weights. The performance of the proposed method was evaluated by the experiments using an ultrasonic imaging phantom. Using the proposed strategies, the lateral full widths at half maximum (FWHM) were both 0.288 mm, which was better than that of 0.348 mm obtained by the mAPES beamformer only. Also, the image contrasts realized by the mAPES beamformer with the PCFs estimated before and after application of the mAPES beamformer weights to the element signals were 5.61 dB and 5.32 dB, respectively, which were better than that of 5.14 dB obtained by the mAPES beamformer only.

**Keywords:** adaptive beamformer; coherence factor; lateral spatial resolution; image contrast

---

## 1. Introduction

Ultrafast ultrasound imaging with parallel beamforming [1] is now frequently used for functional imaging such as the measurement of shear wave propagation [2–4], evaluation of myocardial function [5–7], and blood flow imaging [8–14]. However, the parallel beamforming degrades the lateral spatial resolution and image contrast because unfocused transmit beams are used to illuminate a wide region [2]. The lateral resolution and contrast are important factors determining image quality. Therefore, signal processing methods for improvement of the lateral resolution and contrast are demanded for the ultrafast ultrasound imaging. Spatial compounding [15,16] and synthetic aperture imaging [17,18] have been used as such signal processing methods. However, the spatial compounding and the synthetic aperture imaging require multiple transmissions and, thus, the frame rate is degraded. It would be beneficial if the spatial resolution were improved without compromising the temporal resolution.

Recently, adaptive beamforming has been studied and used in many applications [19,20]. In the delay-and-sum (DAS) beamforming, element echo signals are delayed based on the geometrical information of the focal point and each transducer element. The weights applied to element echo signals in the DAS beamforming are predetermined and independent of the received data. On the other hand, in the adaptive beamformer, weights are adaptively optimized using received echo signals [19]. The minimum variance (MV) beamforming [20], a kind of adaptive beamformer, determines the weights to minimize the power of the beamformer output while maintaining unit gain of a signal of interest. Many researchers have previously attempted to introduce the MV beamformer to the field of the medical ultrasound imaging [21–23], and the MV beamformer provides a significant improvement of the lateral spatial resolution. Also, Synnevåg et al. demonstrated the capability of the MV beamformer to compensate for the degradation of the lateral resolution arising from the parallel beamforming [24]. Blomberg et al. proposed the amplitude and phase estimation (APES) beamformer [25], which eliminates the desired signal from the spatial covariance matrix by DAS beamforming, where the desired signal means the signal which is coherent with the echo from the focal point. We modified the APES beamforming for more accurate estimation of the desired signal by considering the directivity of the array transducer element to discard sub-array averaging [26–29].

On the other hand, adaptive weighting methods based on the coherence factor have also been studied for improvement of the lateral spatial resolution. The coherence factor works as a metric to evaluate the focusing error in receive beamforming, and it is evaluated from echo signals received by individual transducer elements. Li and Li proposed the generalized coherence factor (GCF) [30], which is the ratio of the energy of the low spatial frequency components of element echo signals to the total energy. The direct current (DC) component and the high-frequency components were regarded as the coherent and incoherent signals, respectively and, hence, the GCF represents the degree of the focusing error. Camacho et al. proposed the phase coherence factor (PCF) and the sign coherence factor (SCF) [31], which are evaluated from the phases of the delay compensated echo signals in the DAS beamforming. The PCF is determined by the phase variance among received signals, and the SCF is determined by changes in polarities of the received signals.

For further improvement of the lateral resolution and contrast, in the present study, we examined two strategies to combine the adaptive beamforming and the coherence-based imaging. In one of the two examined strategies, the PCF is estimated using element echo signals before application of the weight vector determined by the adaptive beamformer. In the other strategy, the PCF was evaluated from the element signals subjected to the mAPES beamformer weights.

We also tried to solve a problem in the PCF. Some researchers have already tried to combine the coherence factor with the MV beamforming [32–34]. In those studies, the coherence factor is evaluated from echo signals received by individual transducer elements. However, echoes from a diffuse scattering medium will be suppressed when the PCF is estimated from echo signals received by individual transducer elements because there are many echoes with similar amplitudes and they interfere with each other. To overcome such a problem, we previously proposed the phase coherence imaging with sub-aperture beamforming [35–37]. Sub-aperture beamforming reduces the effect of interference among echoes from diffuse scattering medium, and the visibility of the diffuse scattering medium in phase coherence imaging was improved. In the present study, we also tried to implement the PCF into the modified APES beamformer with sub-aperture beamforming. The performance of the proposed method was evaluated by the experiments using an ultrasound imaging phantom.

## 2. Materials and Methods

### 2.1. Modified Amplitude and Phase Estimation (mAPES) Beamforming [26]

The complex ultrasound signals received by individual transducer elements in an ultrasound array probe are defined as follows:

$$\mathbf{S} = (s_0, s_1, \cdots, s_{M-1})^{\mathrm{T}}, \tag{1}$$

where $^\mathrm{T}$ and $M$ denote the transpose and the number of transducer elements, respectively, and $s_m$ ($m = 0, 1, \cdots, M - 1$) is the complex echo signal received by the $m$-th transducer element. The output signal $u$ is expressed as follows:

$$u = \mathbf{w}^\mathrm{H} \mathbf{S}, \tag{2}$$

where $^\mathrm{H}$ and $\mathbf{w}$ are Hermitian operator and the weight vector applied to the received echo signals, respectively. Let us define the spatial covariance matrix by $\mathbf{R} = \mathrm{E}[\mathbf{S}\mathbf{S}^\mathrm{H}]$, where $\mathrm{E}[\cdot]$ denotes the expectation. In the APES beamforming, the weight $\mathbf{w}_\mathrm{APES}$ is expressed as follows:

$$\mathbf{w}_\mathrm{APES} = \frac{\mathbf{Q}^{-1}\mathbf{a}}{\mathbf{a}^\mathrm{H}\mathbf{Q}^{-1}\mathbf{a}}, \tag{3}$$

where $\mathbf{Q} = \mathbf{R} - \mathbf{G}\mathbf{G}^\mathrm{H}$, $\mathbf{G} = [g_0, g_1, \cdots, g_{M-1}]^\mathrm{T}$, and $\mathbf{a}$ is the steering vector expressed as follows:

$$\mathbf{a} = \begin{pmatrix} a_0 \\ a_1 \\ \vdots \\ a_{M-1} \end{pmatrix} = \begin{pmatrix} \exp\left(-j\frac{2\pi f_0 \sqrt{(x_0 - x_f)^2 - z_f^2}}{c_0}\right) \\ \exp\left(-j\frac{2\pi f_0 \sqrt{(x_1 - x_f)^2 - z_f^2}}{c_0}\right) \\ \vdots \\ \exp\left(-j\frac{2\pi f_0 \sqrt{(x_{M-1} - x_f)^2 - z_f^2}}{c_0}\right) \end{pmatrix}, \tag{4}$$

where $f_0$ and $c_0$ are the ultrasonic center frequency and speed of sound, respectively. The vector $\mathbf{G}$ corresponds to the desired signal from the receiving focal point $(x_f, z_f)$, and $g_m$ ($m = 0, 1, \cdots, M - 1$) is estimated as follows [26]:

$$g_m = \frac{D(\theta_m)\mathbf{a}^\mathrm{H}\mathbf{S}}{\sum_{i=0}^{M-1} D(\theta_i)}, \tag{5}$$

where

$$D(\theta_m) = \frac{\sin\left(\frac{\pi l}{\lambda}\sin\theta_m\right)}{\frac{\pi l}{\lambda}\sin\theta_m}, \tag{6}$$

$$\theta_m = \tan^{-1}\left(\frac{x_f - x_m}{z_f}\right). \tag{7}$$

In our previous study, the outputs of the sub-aperture beamformers were used instead of $\mathbf{S}$ in Equation (2) [26]. The output $y_k$ of the $k$-th sub-aperture ($k = 0, 1, \cdots, K - 1$) consisting of $M_\mathrm{sub} = M/K$ elements is expressed as follows:

$$y_k = \left(a^*_{M_\mathrm{sub}\cdot k}, a^*_{M_\mathrm{sub}\cdot k+1}, \cdots, a^*_{M_\mathrm{sub}\cdot k+M_\mathrm{sub}-1}\right) \begin{pmatrix} s_{M_\mathrm{sub}\cdot k} \\ s_{M_\mathrm{sub}\cdot k+1} \\ \vdots \\ s_{M_\mathrm{sub}\cdot k+M_\mathrm{sub}-1} \end{pmatrix} = \mathbf{a}_k^\mathrm{H}\mathbf{S}_k, \tag{8}$$

$$u_\mathrm{mAPES} = \mathbf{w}_\mathrm{mAPES}^\mathrm{H}\mathbf{Y}, \tag{9}$$

where

$$\mathbf{Y} = (y_0, y_1, \cdots, y_{K-1})^\mathrm{T}, \tag{10}$$

$$\mathbf{w}_\mathrm{mAPES} = \frac{\mathbf{C}^{-1}\mathbf{J}}{\mathbf{J}^\mathrm{H}\mathbf{C}^{-1}\mathbf{J}}, \tag{11}$$

$$\mathbf{C} = \mathbf{YY}^H - \mathbf{VV}^H, \tag{12}$$

$$\mathbf{V} = (v_0, v_1, \cdots, v_{K-1})^T, \tag{13}$$

$$v_k = \frac{b_k \sum_{i=0}^{K-1} y_i}{\sum_{i=0}^{K-1} b_i}, \tag{14}$$

$$b_k = \sum_{i=0}^{M_{\text{sub}}-1} D(\theta_{M_{\text{sub}} \cdot k + i}), \tag{15}$$

where $\mathbf{J}$ is a $K$ dimensional vector of ones.

### 2.2. Gaussian Phase Coherence Factor (gPCF)

The PCF is originally defined by the standard deviation of the phases of echo signals received by individual transducer elements [31] and used for weighting the beamformed echo signals to suppress echoes with focusing errors. In our previous study, the gPCF was proposed to enhance the effect of the PCF [36]. The gPCF is expressed as follows:

$$\text{gPCF} = \exp\left\{\rho \times \left(\frac{\sigma}{\sigma_0}\right)^2\right\}, \tag{16}$$

where $\rho$ is the control parameter, which was set at 3 in the present study [36]. Also, $\sigma_0$ is the nominal standard deviation of $\pi/3^{0.5}$ of the uniform distribution between $-\pi$ and $\pi$ [31], and $\sigma$ is the standard deviation among phases of delay-compensated echo signals received by individual transducer elements or outputs from sub-aperture beamformers. When an echo is coming from the receiving focal point and sound speed in medium is homogeneous, no focusing error occurs and the gPCF is estimated to be 1. On the other hand, when an echo is coming from the out-of-focal point, phase variance increases and the gPCF falls to 0. In our previous study, the outputs of the sub-aperture beamformers were prepared and, then, the phase variance was estimated using outputs of the sub-apertures to suppress interferences from the out-of-focus echoes [35]. In the present study, the gPCF is estimated using the output $y_k$ of each sub-aperture defined in Equation (8) (i.e., output signals before applying the adaptive beamformer weights).

### 2.3. Modifiled APES Beamformer Weighted by gPCF

In the present study, two strategies were examined to implement the gPCF into the mAPES beamformer. In both strategies, the outputs of the mAPES beamforer were weighted by the gPCF, but the gPCFs was estimated differently. The schematic diagrams are shown in Figure 1. In one of the examined strategies, as illustrated in Figure 1a, the gPCF was estimated from the outputs of sub-aperture beamformers before application of the mAPES beamformer weights, i.e., the gPCF is estimated using $y_k$ $(k = 0, 1, \cdots, K-1)$ in Equation (8). In another strategy, the output signals from the sub-aperture beamformer weighted by the mAPES beamfomer weights were used to estimate the gPCF, i.e., the gPCF is estimated using $y_k \cdot w_{\text{mAPES},k}$ $(k = 0, 1, \cdots, K-1)$, where $w_{\text{mAPES},k}$ is the $k$-th element of $\mathbf{w}_{\text{mAPES}}$ defined in Equation (11). To make a B-mode image, either Figure 1a or Figure 1b is adopted to estimate the gPCF applied to the output of the adaptive beamformer. Performances of such two procedures were examined in the subsequent section.

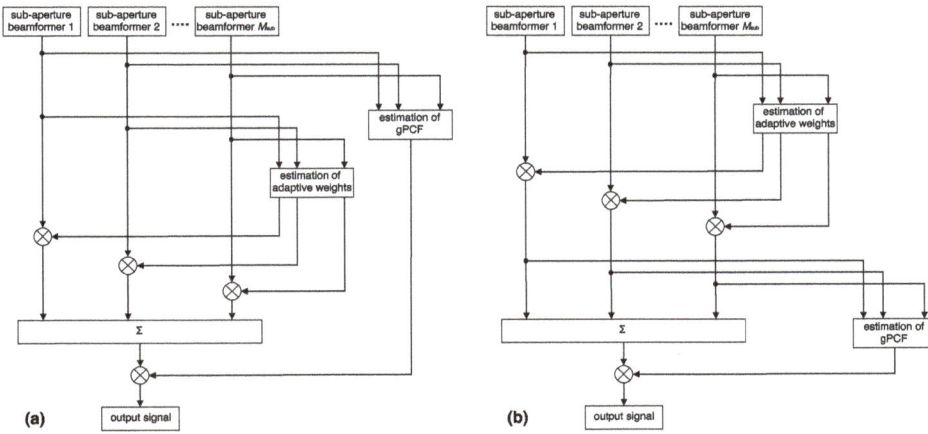

**Figure 1.** Illustration of weighting procedure using Gaussian Phase Coherence Factor (gPCF) estimated (**a**) before and (**b**) after applying adaptive beamformer weights to outputs of sub-aperture beamformers.

## 2.4. Experimental Methods and Evaluation Metrics

### 2.4.1. Experimental Setup

In the present study, an ultrasound imaging phantom (model 040GSE, CIRS, Norfolk, VA, USA) was used for evaluation of image quality. A linear array ultrasonic probe at a nominal center frequency of 7.5 MHz was used. The element pitch of the linear array and the number of the transducer elements were 0.2 mm and 192, respectively. Ultrasonic echo signals received by individual transducer elements were acquired by a custom-made ultrasound scanner (RSYS0002, Microsonic, Tokyo, Japan) at a sampling frequency of 31.25 MHz. The beamforming procedure was performed off-line using the numerical analysis software MATLAB (The MathWorks Inc., Natick, MA, USA). The transmit-receive procedure is described in [9]. In the present study, plane waves were emitted using 96 transducer elements, and then, 24 receiving beams were created in response to one emission. By repeating such a transmit-receive procedure four times, $24 \times 4 = 96$ receiving beams were created at intervals of 0.2 mm. The frame rate achieved under such transmit-receive response condition was 1302 Hz at the pulse repetition frequency of 5208 Hz.

### 2.4.2. Spatial Resolution

The spatial resolution was evaluated using the lateral full width at half maxima obtained from the amplitude profile of an echo from a point scatterer (fine string in the phantom) [26].

### 2.4.3. Contrast

Image contrast $C$ was evaluated as follows [26]:

$$C = \frac{|\mu_b - \mu_l|}{(\mu_b + \mu_l)/2},$$

(17)

where $\mu_b$ and $\mu_l$ are mean gray levels in background and lesion, respectively. In the present study, an anechoic cyst phantom was adopted as the lesion, and a diffuse scattering medium was adopted as the background.

### 2.4.4. Peak-to-Speckle Ratio

In the present study, the peak-to-speckle ratio is defined as the ratio of a peak gray level at a strong scatterer (fine string in the phantom) to the mean gray level in diffuse scattering medium.

## 3. Results

### 3.1. Basic Experimental Results Using Phantom

Figure 2a–d show B-mode images of a string phantom obtained by the conventional DAS beamforming, the mAPES beamforming without gPCF, and those with gPCFs estimated before and after applying the adaptive beamformer weights, respectively.

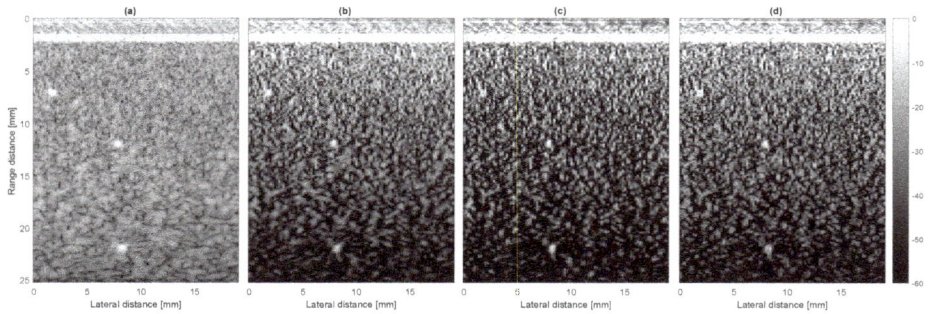

**Figure 2.** B-mode images of string phantom obtained (**a**) with delay-and-sum (DAS); (**b**) with modified amplitude and phase estimation (mAPES) beamforming without gPCF, and mAPES beamforming with gPCFs evaluated (**c**) before and (**d**) after applying the adaptive beamformer weights.

In the phantom used for this experiment, three point targets were placed at different axial depths. Figure 3a,b show the lateral amplitude profiles with respect to point targets at axial depths of 12 mm and 22 mm in Figure 2a–d.

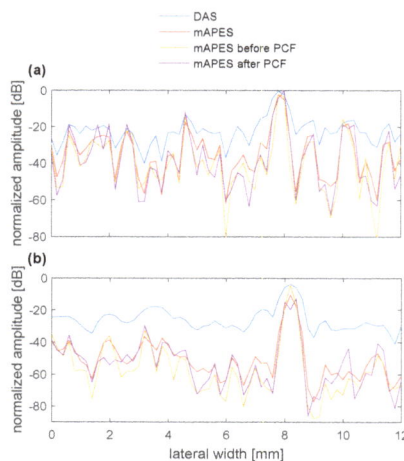

**Figure 3.** Lateral amplitude profile obtained from point target at axial depths of (**a**) 12 mm and (**b**) 22 mm in Figure 2.

In Figure 3a,b, the lateral amplitude profile obtained with the proposed method was sharpened and speckles obtained with the proposed method were resolved as compared to those obtained with the conventional mAPES beamformer. The lateral full widths at half maxima of the lateral amplitude profiles shown in Figure 3a obtained by DAS, mAPES without gPCF, and those with gPCFs estimated before and after applying the adaptive beamformer weights were 0.560, 0.392, 0.356, and 0.344 mm, respectively. Also, the lateral full widths at half maxima of the lateral amplitude profiles shown in Figure 3b obtained by DAS, mAPES without gPCF, and those with gPCFs estimated before and after applying the adaptive beamformer weights were 0.668, 0.348, 0.288, and 0.288 mm, respectively. Figure 4a,b show that the phases of the outputs of the sub-aperture beamformers before and after applying the adaptive beamformer weights, respectively, which was obtained at range and lateral positions of 22 mm and 3.60 mm, respectively.

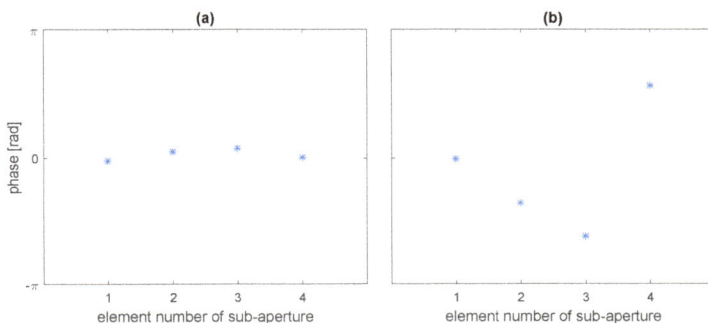

**Figure 4.** Phase of the signal obtained by each sub-aperture element (**a**) before and (**b**) after applying the adaptive weights.

These results show that the phases shown in Figure 4b deviate among the sub-apertures, whereas those shown in Figure 4a were well aligned.

Figure 5a–d show B-mode images of a cyst phantom obtained by the DAS beamforming, the mAPES beamforming without the gPCF, and those with the gPCFs estimated before and after applying the adaptive beamformer weights, respectively.

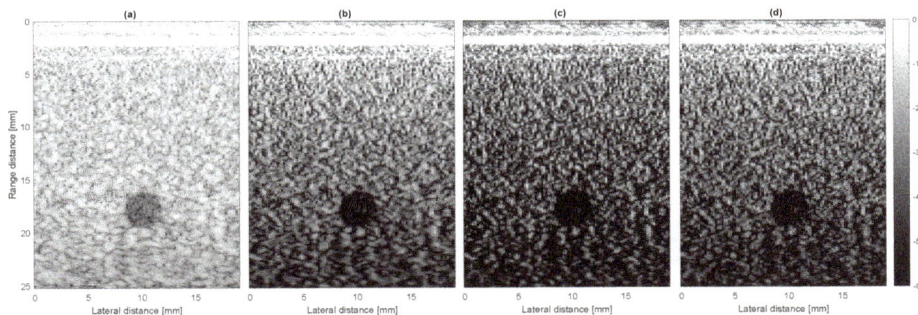

**Figure 5.** B-mode images of cyst phantom obtained (**a**) with DAS; (**b**) with mAPES beamforming without gPCF, and mAPES beamforming with gPCFs evaluated (**c**) before and (**d**) after applying the beamformer weights.

In this experiment, an anechoic cyst was embedded in the phantom. A rectangular region (from 16.8 to 19.2 mm in the range direction, from 9 to 11 mm in the lateral direction) was regarded as the lesion, and another rectangular region (from 16.8 to 19.2 mm in the range direction, from 5 to 7 mm in the lateral direction) was regarded as background. Contrasts of the B-mode image obtained by DAS, mAPES without gPCF, and those with gPCFs estimated before and after applying the adaptive beamformer weights were 4.05, 5.14, 5.61, and 5.32 dB, respectively. Experimental results show that image contrast was improved by the proposed method.

Figure 6a,b show B-mode images of the wire phantom and the cyst phantom obtained by the adaptive beamforming combined with the PCF estimated without the sub-aperture beamforming.

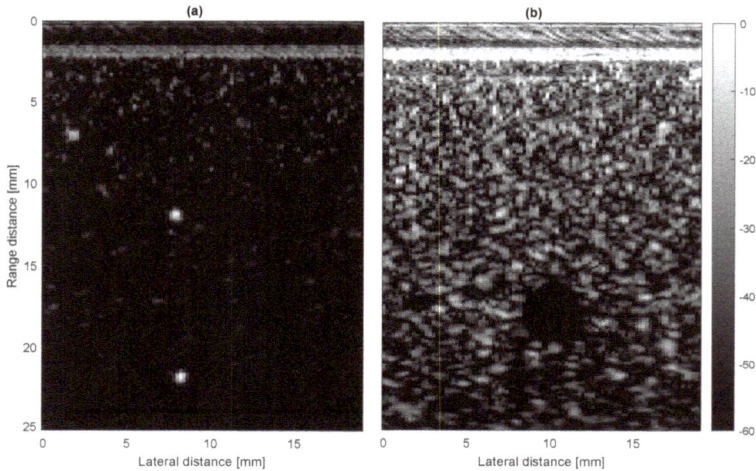

**Figure 6.** B-mode images of (**a**) wire phantom and (**b**) cyst phantom obtained by mAPES beamformer with gPCF estimated in a conventional way, i.e., gPCF was estimated without sub-aperture beamforming. Element echo signals before application of mAPES beamformer weights were used for estimation of gPCF.

In Figure 6a,b, the gPCF was estimated before applying the adaptive beamforming weights. The lateral full widths at half maxima of point targets at depths of 12 mm and 22 mm in Figure 6a were 0.356 mm and 0.280 mm, respectively, and the contrast of the B-mode image in Figure 6b was estimated to be 5.99 dB. As shown in Figure 6a, echoes from diffuse scattering medium are suppressed by the PCF when sub-aperture beamforming was not used, and such a negative effect of the PCF estimated from element echo signals can be reduced by sub-aperture beamforming as shown in Figure 3. Figure 7a,b show the lateral resolution and the peak-to-speckle ratio at each number of sub-apertures obtained from point targets at axial depths of 12 mm and 22 mm, respectively. In the present study, mean gray levels in a rectangular region (from 10.8 to 13.2 mm in the range direction, from 3 to 5 mm in the lateral direction) and another rectangular region (from 20.8 to 23.2 mm in the range direction, from 3 to 5 mm in the lateral direction) were used for evaluation of the peak-to-speckle ratios in Figure 7a,b, respectively. The gPCF obtained without sub-aperture beamforming corresponds to the number of sub-apertures of 72.

In Figure 7a,b, the lateral full widths at half maxima do not vary with increasing the number of sub-apertures, whereas the peak-to-speckle ratios increase. Experimental results show that the gray level in diffuse scattering medium becomes relatively low under a large number of sub-apertures and, hence, the speckles are not visualized well when the proposed method is used without sub-aperture beamforming.

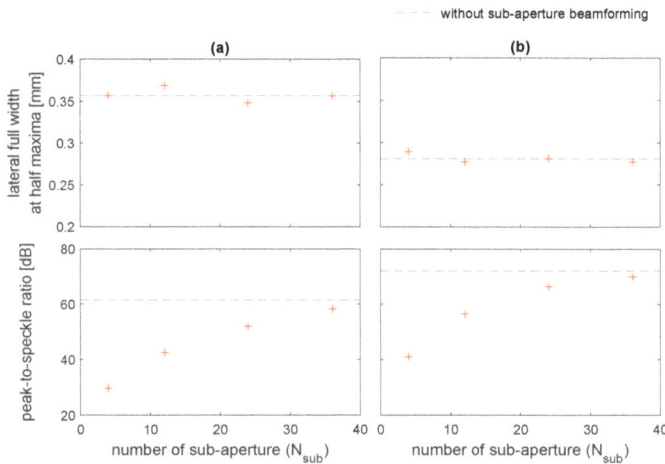

**Figure 7.** Changes in (**top**) lateral full widths at half maxima and (**bottom**) peak-to-speckle ratio obtained by the proposed method with gPCF estimated before applying the adaptive beamformer weights. Results were obtained with respect to point targets at axial depths of (**a**) 12 mm and (**b**) 22 mm.

### 3.2. In Vivo Measurement of Human Carotid Artery

In the present study, the feasibility of the proposed method under in vivo condition was evaluated by measurement of a human carotid artery. Figure 8a–d show B-mode images of the carotid artery obtained by the DAS beamforming, the mAPES beamforming without gPCF, and those with gPCFs estimated before and after applying the adaptive beamformer weights, respectively.

In Figure 8c,d, the wall of the carotid artery was depicted more clearly and sharpened along the lateral direction than Figure 8a,b.

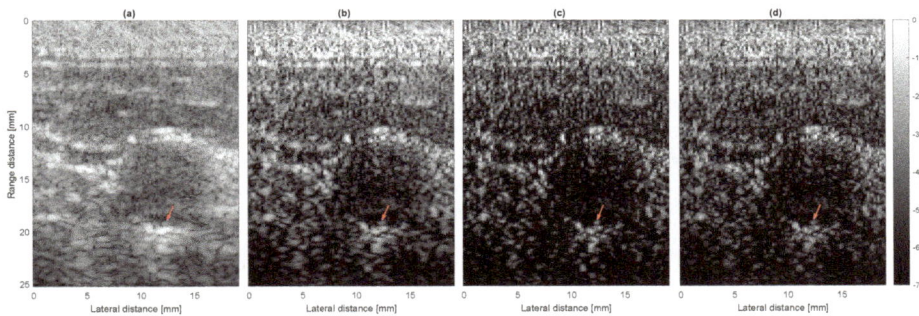

**Figure 8.** B-mode images of carotid artery obtained (**a**) with DAS; (**b**) with mAPES beamforming without gPCF, and those with gPCFs estimated (**c**) before and (**d**) after applying the beamformer weights.

## 4. Discussion

In the present study, we tried to combine the modified adaptive beamformer with the phase coherence weighting. From the experiments using a wire phantom, the lateral full width at half maxima obtained by the proposed method was better than conventional DAS beamforming and adaptive beamforming only. Also, in a B-mode image shown in Figure 2c, the speckles in diffuse scattering medium in the axial deep region were more clearly resolved than in Figure 2b. Furthermore, as shown in Figures 2c and 6a, obtained results show that the visibility of echoes from diffuse scattering medium was significantly improved by applying the gPCF to the outputs of the sub-aperture beamformers.

In the present study, we also tried to estimate the PCF using the phase variance of the outputs of the sub-aperture beamformers, which were subjected to the weight vector determined by the mAPES beamformer. In such a case, as shown in the B-mode image in Figure 2d, the spatial resolution became worse as compared to that shown in Figure 2c, in which the PCF was estimated from the outputs of the sub-aperture beamformers before applying the adaptive weights. The mAPES beamformer changes the amplitude and phase of the outputs of the sub-aperture beamformers adaptively so that the lateral resolution is optimized. In Figure 4a,b, obtained results show that the phase rotation was induced by the adaptive beamformer weights even with the echo signals from a strong scatterer and, therefore, the gPCF became low. With respect to the point target at an axial depth of 12 mm, a point scatterer depicted in Figure 2d was blurred as compared to that depicted in Figure 2c. This is because the peak value of the echo from the point scatterer decreased by the gPCF from echo signals after adaptive weighting but the phase variance of the signals other than the peak position is originally large and adaptive weights do not change the gPCF value significantly.

From the experimental results on a cyst phantom, which are shown in Figure 5a–d, image contrast was improved using the gPCF estimated before and after applying the adaptive beamformer weights. Both the adaptive beamformer and the PCF were proposed for improvement of the spatial resolution but improvement of contrast was accomplished by applying the gPCF to the output of the modified adaptive beamformer. From the experimental results on two phantoms, the gPCF evaluated using echo signals subjected to mAPES beamformer weights was shown to give worse results than that evaluated using echo signals without mAPES beamformer weights. Consequently, evaluating the gPCF using echo signals subjected to adaptive beamformer weights is not preferred.

In the B-mode images in Figure 2, a point target at a depth of about 7 mm is somewhat blurred also in the B-mode image obtained by DAS beamforming (Figure 2a). In the near field, the ultrasound field fluctuates significantly due to interference among ultrasonic waves emitted from the transducer [38], and the performance of image formation is degraded. The performance of the proposed method could be degraded because such interference would also affect the proposed method, which uses the amplitude and phase information of the measured ultrasonic signal. Therefore, we need more improvements in the beamformation process for near-field imaging.

In the proposed method, the penetration depths in the obtained B-mode images were about 30 mm because the center frequency of the linear array probe used in the present study was 7.5 MHz and it was fabricated for imaging of superficial tissues. To increase the penetration depth up to 10–20 cm, we need to use an ultrasonic probe at a lower center frequency. The performance of the proposed method on such probes at lower center frequencies will be investigated in our future work.

In the present study, the B-mode images obtained without the sub-aperture beamforming were also evaluated as the previously proposed method. With respect to the lateral resolution, the appearance of the fine wire depicted in Figure 6a does not seem to change as compared to Figure 2c. On the other hand, the speckle visibility in the obtained B-mode image was degraded. In ultrafast ultrasound imaging, the adaptive beamformer incorporating with the coherence factor degrades the speckle visibility in the obtained image [39]. In the present study, the output signals from the sub-aperture beamformers were used in estimation of both the beamformer weights and the gPCF, and experimental results obtained with sub-aperture beamforming show that such an implementation is effective to avoid degradation of the speckle visibility while maintaining the improvement of the spatial resolution. With respect to image contrast, better contrast was obtained by the proposed method without sub-aperture beamforming. However, when the proposed method is performed without sub-aperture beamforming, the dimension of the covariance matrix becomes large and high computational cost is required to calculate the inverse matrix of the covariance matrix. The sub-aperture beamforming can reduce the dimension of the covariance matrix, and accordingly, the computation time becomes shorter [27].

As described above, although the proposed method needs to carry out an additional processing for weighting the coherence factor on the conventional modified adaptive beamforming, the gPCF estimated with sub-aperture beamforming is a computationally effective method. In the present study, better lateral resolution and contrast were realized with slightly increased the processing time compared to the mAPES beamformer only. The computation time of the proposed method was 339% of the conventional DAS beamformer, when an Intel(R) Core(TM) i5-6500 CPU was used with 8 GB of RAM. The proposed method has potential to be used in practical applications by incorporating parallel processing techniques with the proposed method.

## 5. Conclusions

The spatial resolution and image contrast are important factors determining the quality of an ultrasound B-mode image. In the present study, we introduced the modified adaptive beamformer enhanced by PCF for further improvement of image quality. From the results of the phantom experiments, the spatial resolution evaluated by the lateral full widths at half maxima of echoes from point targets and image contrast were improved by the proposed method.

**Acknowledgments:** This study was supported by JSPS KAKENHI Grant Numbers JP17H03276 and JP15K13995.

**Author Contributions:** Hideyuki Hasegawa conceived and designed the experiments. Michiya Mozumi analyzed the data. Michiya Mozumi and Hideyuki Hasegawa wrote the paper.

**Conflicts of Interest:** The authors declare no conflict of interest.

## References

1. Shattuck, D.P.; Weinshenker, M.D. Explososcan: A parallel processing technique for high speed ultrasound imaging with linear phase array. *J. Acoust. Soc. Am.* **1984**, *75*, 1273–1282.
2. Tanter, M.; Fink, M. Ultrafast imaging in biomedical ultrasound. *IEEE Trans. Ultrason. Ferroelectr. Freq. Control* **2014**, *61*, 102–119.
3. Tanter, M.; Bercoff, J.; Sandrin, L.; Fink, M. Ultrafast compounding imaging for 2-D motion vector estimation: application to transient elastography. *IEEE Trans. Ultrason. Ferroelectr. Freq. Control* **2002**, *49*, 1363–1374.
4. Bercoff, J.; Tanter, M.; Fink, M. Supersonic shear imaging: A new technique for soft tissue elasticity mapping. *IEEE Trans. Ultrason. Ferroelectr. Freq. Control* **2004**, *51*, 396–409.
5. Honjo, Y.; Hasegawa, H.; Kanai, H. Accurate ultrasonic measurement of myocardial regional strain rate at high temporal and spatial resolutions. In Proceedings of the 2008 IEEE International Ultrasonics Symposium (IUS), Beijing, China, 2–5 November 2008; pp. 1995–1998.
6. Provost, J.; Nguyen, V.T.-H.; Legrand, D.; Okrasinski, S.; Costet, A.; Gambhir, A.; Garan, H.; Konofagou, E.E. Electromechanical wave imaging for arrhythmias. *Phys. Med. Biol.* **2011**, *56*, L1–L11.
7. Cikes, M.; Tong, L.; Sutherland, G.R.; D'hooge, J. Ultrafast cardiac ultrasound imaging: Technical principles, applications, and clinical benefits. *JACC Cardiovasc. Imaging* **2014**, *7*, 812–823.
8. Udesen, J.; Gran, F.; Hansen, K.L.; Jensen, J.A.; Thomsen, C.; Nielsen M.B. High frame-rate blood vector velocity imaging using plane waves: Simulations and preliminary experiments. *IEEE Trans. Ultrason. Ferroelectr. Freq. Control* **2008**, *55*, 1729–1743.
9. Hasegawa, H.; Kanai, H. Simultaneous imaging of artery wall strain and blood flow by high frame rate acquisition of RF signals. *IEEE Trans. Ultrason. Ferroelectr. Freq. Control* **2008**, *55*, 2626–2639.
10. Bercoff, J.; Montaldo, G.; Loupas, T.; Savery, D.; Mézière, F.; Fink, M.; Tanter, M. Ultrafast compound Doppler imaging: providing full blood flow characterization. *IEEE Trans. Ultrason. Ferroelectr. Freq. Control* **2011**, *58*, 134–147.
11. Yiu, B.Y.; Yu, A.C. High-frame-rate ultrasound colorencoded speckle imaging of complex flow dynamics. *Ultrasound Med. Biol.* **2013**, *39*, 1015–1025.
12. Takahashi, H.; Hasegawa, H.; Kanai, H. Temporal averaging of two-dimensional correlation functions for velocity vector imaging of cardiac blood flow. *J. Med. Ultrason.* **2015**, *42*, 323–330.
13. Takahashi, H.; Hasegawa, H.; Kanai, H. Echo motion imaging with adaptive clutter filter for assessment of cardiac blood flow. *Jpn. J. Appl. Phys.* **2015**, *54*, doi:10.7567/JJAP.54.07HF09.

14. Jensen, J.A.; Nikolov, S.I.; Yu, A.C.H.; Garcia, D. Ultrasound vector flow imaging–Part II: Parallel system. *IEEE Trans. Ultrason. Ferroelectr. Freq. Control* **2016**, *63*, 17221732.

15. Montaldo, G.; Tanter, M.; Bercoff, J.; Benech, N. Fink, M. Coherent plane-wave compounding for very high frame rate ultrasonography and transient elastography. *IEEE Trans. Ultrason. Ferroelectr. Freq. Control* **2009**, *56*, 489–506.

16. Denarie, B.; Tangen, T.A.; Ekroll, I.K.; Rolim, N.; Torp, H.; Bjåstad, T.; Lovstakken, L. Coherent plane-wave compounding for very high frame rate ultrasonography of rapidly moving targets. *IEEE Trans. Med. Imaging* **2013**, *32*, 1265–1276.

17. Jensen, J.A.; Nikolov, S.I.; Gammelmark, K.L.; Pedersen, M.H. Synthetic aperture ultrasound imaging. *Ultrasonics* **2006**, *44*, e5–e15.

18. Hasegawa, H.; De Korte, C.L. Impact of element pitch on synthetic aperture ultrasound imaging. *J. Med. Ultrason.* **2016**, *43*, 317–325.

19. Veen, B.D.V.; Buckley, K.M. Beamforming: A versatile approach to spatial filtering. *IEEE ASSP Mag.* **1988**, *5*, 4–24.

20. Capon, J. High-resolution frequency-wavenumber spectrum analysis. *Proc. IEEE* **1969**, *57*, 1408–1418.

21. Sasso, M.; Cohen-Bacrie, C. Medical ultrasound imaging using the fully adaptive beamformer. In Proceedings of the IEEE International Conference on Acoustics, Speech, and Signal Processing, Philadelphia, PA, USA, 23 March 2005; pp. 489–492.

22. Synnevåg, J.F.; Austeng, A.; Holm, S. Adaptive beamforming applied to medical ultrasound imaging. *IEEE Trans. Ultrason. Ferroelectr. Freq. Control* **2007**, *54*, 1606–1613.

23. Holfort, I.K.; Gran, F.; Jensen, J.A. Broadband minimum variance beamforming for ultrasound imaging. *IEEE Trans. Ultrason. Ferroelectr. Freq. Control* **2009**, *56*, 314–325.

24. Synnevåg, J.F.; Austeng, A.; Holm, S. Benefits of minimum-variance beamforming in medical ultrasound imaging. *IEEE Trans. Ultrason. Ferroelectr. Freq. Control* **2009**, *56*, 1868–1879.

25. Blomberg, A.E.A.; Holfort, I.K.; Austeng, A.; Synnevåg, J.F.; Jensen, J.A. APES beamforming applied to the ultrasound imaging. In Proceedings of the 2009 IEEE International Ultrasonics Symposium (IUS), Rome, Italy, 20–23 September 2009; pp. 2347–2350.

26. Hasegawa, H.; Kanai, H. Effect of element directivity on adaptive beamforming applied to high-frame-rate ultrasound. *IEEE Trans. Ultrason. Ferroelectr. Freq. Control* **2015**, *62*, 511–523.

27. Hasegawa, H. Improvement of penetration of modified amplitude and phase estimation beamformer. *J. Med. Ultrason.* **2017**, *44*, 3–11.

28. Hasegawa, H. Apodized adaptive beamformer. *J. Med. Ultrason.* **2017**, *44*, 155–165.

29. Hasegawa, H. Adaptive beamforming applied to transverse oscillation. In Proceedings of the 2017 IEEE International Ultrasonics Symposium (IUS), Washington, DC, USA, 6–9 September 2017; pp. 1–4.

30. Li, P.-C.; Li, M.-L. Adaptive imaging using the generalized coherence factor. *IEEE Trans. Ultrason. Ferroelectr. Freq. Control* **2003**, *50*, 128–141.

31. Camacho, J.; Parrilla, M.; Fritsch, C. Phase coherence imaging. *IEEE Trans. Ultrason. Ferroelectr. Freq. Control* **2009**, *56*, 958–974.

32. Asl, B.M.; Mahloojifar, A. Minimum variance beamforming combined with adaptive coherence weighting applied to medical ultrasound imaging. *IEEE Trans. Ultrasonics. Ferroelectr. Freq. Control* **2009**, *56*, 1923–1931.

33. Nilsen, C.-I.C.; Holm, S. Wiener beamforming and the coherence factor in ultrasound imaging. *IEEE Trans. Ultrasonics. Ferroelectr. Freq. Control* **2010**, *57*, 1329–1346.

34. Chen, H.; Li, M. Improved high axial resolution ultrasound imaging using spectral whitening and minimum-variance based coherence weighting. In Proceedings of the 2017 IEEE International Ultrasonics Symposium (IUS), Washington, DC, USA, 6–9 September 2017; pp. 1–4.

35. Hasegawa, H.; Kanai, H. Effect of subaperture beamforming on phase coherence imaging. *IEEE Trans. Ultrason. Ferroelectr. Freq. Control* **2014**, *61*, 1779–1790.

36. Hasegawa, H. Enhancing effect of phase coherence factor for improvement of spatial resolution in ultrasonic imaging. *J. Med. Ultrason.* **2016**, *43*, 19–27.

37. Fujita H.; Hasegawa, H. Effect of frequency characteristic of excitation pulse on lateral spatial resolution in coded ultrasound imaging. *Jpn. J. Appl. Phys.* **2017**, *56*, doi:10.7567/JJAP.56.07JF16.

38. Franco, E.E.; Andrade, M.A.B.; Adamowski, J.C.; Buiochi, F. Acoustic beam modeling of ultrasonic transducers and arrays using the impulse response and the discrete representation methods. *J. Braz. Soc. Mech. Sci. Eng.* **2011**, *33*, 408–416.

39. Varray F.; Kalkhoran M.A.; Vray D. Adaptive minimum variance coupled with sign and phase coherence factor in IQ domain for plane wave beamforming. In Proceedings of the 2016 IEEE International Ultrasonics Symposium (IUS), Tours, France, 18–21 September 2016; pp. 1–4.

![applied sciences logo] *applied sciences*

MDPI

*Article*

# A PSF-Shape-Based Beamforming Strategy for Robust 2D Motion Estimation in Ultrafast Data

Anne E. C. M. Saris [1,*], Stein Fekkes [1], Maartje M. Nillesen [1], Hendrik H. G. Hansen [1] and Chris L. de Korte [1,2]

[1] Medical Ultra Sound Imaging Center, Department of Radiology and Nuclear Medicine, Radboud University Medical Center, P.O. Box 9101, 6500 HB Nijmegen, The Netherlands; Stein.Fekkes@radboudumc.nl (S.F.); m.nillesen@gmail.com (M.M.N.); Rik.Hansen@radboudumc.nl (H.H.G.H.); Chris.deKorte@radboudumc.nl (C.L.d.K.)
[2] Physics of Fluids Group, MIRA Institute for Biomedical Technology and Technical Medicine, University of Twente, P.O. Box 217, 7500 AE Enschede, The Netherlands
* Correspondence: Anne.Saris@radboudumc.nl; Tel.: +31-243-614-730

Received: 15 February 2018; Accepted: 9 March 2018; Published: 13 March 2018

**Abstract:** This paper presents a framework for motion estimation in ultrafast ultrasound data. It describes a novel approach for determining the sampling grid for ultrafast data based on the system's point-spread-function (PSF). As a consequence, the cross-correlation functions (CCF) used in the speckle tracking (ST) algorithm will have circular-shaped peaks, which can be interpolated using a 2D interpolation method to estimate subsample displacements. Carotid artery wall motion and parabolic blood flow simulations together with rotating disk experiments using a Verasonics Vantage 256 are used for performance evaluation. Zero-degree plane wave data were acquired using an ATL L5-12 ($f_c$ = 9 MHz) transducer for a range of pulse repetition frequencies (PRFs), resulting in 0–600 μm inter-frame displacements. The proposed methodology was compared to data beamformed on a conventionally spaced grid, combined with the commonly used 1D parabolic interpolation. The PSF-shape-based beamforming grid combined with 2D cubic interpolation showed the most accurate and stable performance with respect to the full range of inter-frame displacements, both for the assessment of blood flow and vessel wall dynamics. The proposed methodology can be used as a protocolled way to beamform ultrafast data and obtain accurate estimates of tissue motion.

**Keywords:** ultrafast imaging; beamforming; motion estimation; plane wave; speckle tracking; speckle characteristics; carotid artery

---

## 1. Introduction

Ultrasound imaging is often used to visualize and estimate motion in our body [1–10]. Tissue motion and deformation can provide information about the structure, composition and functioning of the tissue and changes in the dynamic behavior of tissue can be used as indicator of disease. For example, stiff regions in breasts are often related to cancers [11,12], the assessment of local arterial deformation provides information about the atherosclerotic progression and vulnerability of lesions [13–16], increased blood velocities indicate the presence of a stenosis in the vessel [17–19], and infarcted myocardial tissue shows changed deformation patterns as compared to healthy tissue [20].

Speckle tracking (ST), first introduced by Trahey et al. [21,22], is one of the motion estimation methods which is frequently utilized. Speckle patterns are tracked to find the displacement of the underlying tissue. Pattern matching techniques are used, where a kernel region in the first acquisition is matched within a surrounding search region in the following acquisition. The location of the best match defines the displacement of the kernel region and thus the displacement of the underlying tissue. ST can be performed using radio frequency (RF) signals [23], envelope signals [24], B-mode images [20,22], or a combination thereof [25]. Bohs et al. [26] wrote a review on the development of this technique, focusing on multi-dimensional flow estimation. They concluded that decorrelation of speckle patterns is one of the major challenges in ST. Strong displacement gradients and velocities with low beam-to-flow angles were described as major sources of speckle decorrelation. Increasing the imaging frame rate can minimize speckle decorrelation [26].

The first published approach for increasing imaging frame rate was by reconstructing multiple image lines simultaneously from each transmit, so-called parallel beamforming [27]. Nowadays, plane wave and diverging wave imaging are used more often, where the transmitted pulse fully illuminates the entire region of interest, enabling full image reconstruction for every transmit event. This allows for imaging at ultrafast frame rates in the kHz range. Hereby, decorrelation of speckle can be minimized and fast moving, sometimes short-lived, motion patterns can be tracked [28–32]. However, frame-to-frame tissue displacements become smaller, since they scale with the imaging frame rate. Consequently, acquiring accurate subsample displacement estimates is of great importance.

The estimation of subsample displacements in ST was already a challenge before the introduction of ultrafast data acquisition, since echo signals are sampled. For conventionally acquired, line-by-line data, interpolation of the pattern matching function is often applied to acquire subsample displacement estimates [25,33,34]. The sampling grid for conventional data is predefined, with normally a line distance equal to the pitch and an axial grid spacing with typically four samples per wavelength. Consequently, the sampling of the pattern matching function is also set. A predefined shape or curve, often parabolic or cosine [34,35], is assumed and fitted to the peak of the function, either in each direction of the 2D function independently [25,35], or by fitting in 2D to obtain a joint estimate of subsample displacement in both directions. Zahiri Azar et al. [33] investigated 2D polynomial fitting to get a joint estimation of subsample displacement and showed significant improvements in terms of bias and standard deviation (SD) compared to the commonly used 1D parabolic and cosine fitting. Interpolating in between echo signals is also often applied to reduce the error of the subsample displacement estimates [25,36,37].

With the use of ultrafast data, subsample displacements estimation remains a challenge. However, it also offers a totally new possibility: since image reconstruction is performed after receiving the raw channel data, the data sampling grid is not fixed and focusing can be performed at each specific location. Redefining the data sampling grid directly influences the sampling of the pattern matching function. Hence, it allows to choose the available samples and their spacing throughout the peak of the pattern matching function, which we will refer to as the appearance or shape of the function throughout this paper. The possibility to influence its appearance allows to optimize the match between the peak of the function and the interpolation method. This will in theory enable more accurate subsample displacement estimation [34].

More work has been published recently on the combination of ultrafast imaging and ST, for the assessment of blood velocity [28,29,38] and vessel wall strain [39–42]. Here, subsample resolution displacement estimates were obtained by performing 1D [28,29,38,43] or 2D [42] parabolic interpolation of the pattern matching function. No clear reasoning was provided concerning the match between this interpolation function and the sampling, or shape, of the pattern matching function, which is a direct result of the spacing in the ultrasound sampling grid.

In this work, we propose a new concept that combines the choice of the beamforming grid with the choice of the subsample estimator. By doing so, the available data and interpolator can be matched. For this, the sampling grid for ultrafast data was based on the dimensions of the point-spread-function (PSF) of the imaging system. When ST is performed, the pattern matching function is hereby captured in a standardized way, with equal amount of information, i.e., sample points, in the axial and lateral direction throughout the peak. The use of a 2D cubic subsample interpolator, which matches with the shape of the peak, allows for a joint estimate of subsample displacement in both directions. We hypothesize that this will increase the accuracy of subsample displacement estimation and thereby will enlarge the range of detectable displacements and velocities. The concept of this technique and first simulation results on parabolic blood flow were presented in a conference proceeding [44]. The present paper fully describes the reasoning and theory behind the method and presents in vitro rotating disk experiments and carotid artery vessel wall simulations to further study the performance of the method for both vector velocity imaging and strain estimation in the carotid artery. Both wall and flow dynamics were studied, showing a wide range of inter-frame displacements and displacement gradients, to investigate the applicability of the proposed method in a broad field.

## 2. Materials and Methods

### 2.1. Theory

#### 2.1.1. Multi-Step Speckle Tracking

Tissue displacements were estimated using a multi-step ST algorithm. Such multi-scale methods are able to obtain accurate displacement estimates in highly discontinuous displacement fields [45,46]. Normalized 2D cross-correlation was used as the pattern matching function. Inter-frame displacements were first estimated using relatively large 2D windows of envelope (demodulated RF) data, where after, in the second step of the algorithm, smaller windows of RF data were used to refine the displacements. Subsample displacement estimates were obtained by performing interpolation of the cross-correlation-function (CCF).

#### 2.1.2. PSF-Shape-Based Beamforming Grid

The shape of the peak of the CCF is related the PSF of the imaging system, i.e., the combination of transducer choice and transmit sequence. Together with the data sampling grid, the PSF strongly influences the appearance of the peak in the CCF. By incorporating knowledge about the PSF dimensions into the data sampling grid, the CCFs can be captured in a standardized way, with an equal number of samples in the axial and lateral direction throughout the peak. These so-called circularly shaped peaks, i.e., peaks with circular isocontours, provide an equal amount of sample points to the subsample estimator in both directions. This ensures the interpolator gives equal weight to the available information in both directions. Using a 2D interpolator, which matches with these circular shaped peaks, allows for joint 2D subsample displacement estimation. In this work, we investigate whether standardization of the shape, or sampling, of the peak combined with 2D cubic interpolation increases the accuracy and precision of multi-step ST-based displacement estimation.

Circular-shaped CCF peaks were realized by choosing the data sampling grid such that a single period of the system's PSF contained an equal amount of sampling points in axial and lateral direction. Firstly, concerning the RF data (Figure 1a), six samples per wavelength were beamformed in axial direction, in accordance with the all-positive-samples condition [34], resulting in an axial sampling distance of $d_{ax,rf} = c/(2 \times f_c \times 6)$, with c the speed of sound and $f_c$ the center frequency of the emitted waveform. The lateral sampling distance was determined according to $d_{lat,rf} = d_{ax,rf} \times PSF_{ratio} \times N_{cycles}$, with $PSF_{ratio}$ the average ratio between the lateral and axial PSF dimensions and $N_{cycles}$ the number of cycles present in the PSF axially. The envelope data sampling grid (Figure 1b) was generated by down sampling the RF sampling grid in axial direction with the factor $N_{cycles}$, resulting in an axial spacing of $d_{ax,env} = d_{ax,rf} \times N_{cycles}$. The lateral spacing for the envelope data was set similar to the lateral spacing of the RF data ($d_{lat,env} = d_{lat,rf}$). With these spacing settings, the peak of the CCF will appear circular-shaped when using both envelope and RF data in the multi-step ST algorithm.

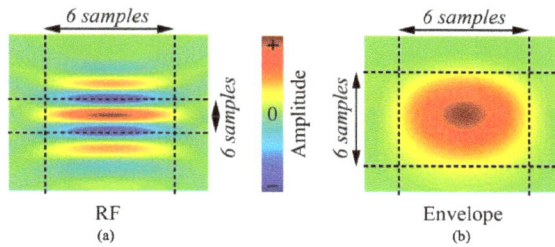

**Figure 1.** Representation of the system's point-spread-function (PSF) visualized using radio frequency (RF) (**a**) and envelope (**b**) data. The data sampling points required to capture a circular-shaped peak in the cross-correlation function are indicated in the figure. Please note that the envelope data sampling grid is a downsampled version of the RF data sampling grid.

## 2.2. Imaging Setup

An L5-12 ATL (ATL, Bothell, WA, USA), 38 mm linear array transducer was used in this work. Zero-degree plane waves were transmitted by simultaneously activating 128 elements at the center of the transducer, using a rectangular apodization window. The same 128 elements were also active in receive. The transducer parameters are listed in Table 1. The Field II ultrasound simulation program [47,48] was used for simulations and the Verasonics Vantage 256 experimental ultrasound system (Verasonics, Kirkland, WA, USA) for the experimental measurements.

**Table 1.** Transducer properties.

| Properties L12-5, 38 mm | Experiments | Simulations |
|---|---|---|
| Center frequency ($f_c$) | 8.9 MHz | 9 MHz |
| Sampling frequency | 36 MHz | 180 MHz |
| Transducer element pitch | 198 µm | |
| Transducer element height | 4 mm | |
| Transducer element width | 173 µm | |
| Nr. of transducers elements | 192 | |
| Nr. of active elements | 128 | |
| Excitation signal | 3-cycle pulse at $f_c$ | |

## 2.3. Simulations

Ultrasound RF element data of blood and tissue were simulated, which are represented by a set of random point scatterers. About 300 scatterers per mm$^3$ where used and each physical transducer element was subdivided into 10 by 10 mathematical elements to increase computational accuracy. A sampling frequency of 180 MHz was used, the resulting RF element data were down sampled to 36 MHz to mimic experimental settings. The speed of sound was assumed to be 1540 m/s.

### 2.3.1. Parabolic Flow in Vessel Phantom

Parabolic flow through a rigid, straight tube with a peak velocity ($v_{max}$) of 1.5 m/s was simulated at flow angles of 75° till 105°, with steps of 3°. The vessel had a lumen diameter of 6 mm and was centered at 20 mm depth. Scatterers, representing the blood, had Gaussian distributed amplitudes (mean = 5, SD = 1). Plane wave ultrasound data were simulated for pulse repetition frequencies (PRFs) of 2.5 till 15 kHz, with steps of 2.5 kHz. Table 2 provides an overview of the simulated PRFs and resulting inter-frame displacements. For each unique combination of flow angle and PRF, a pre and post frame were simulated, where in between the scatterers were propagated according to the parabolic velocity profile. This process was repeated ten times, with random initial scatterer positions, allowing 10 independent velocity estimates. No clutter filtering or time averaging were performed.

Band limited noise was added to the beamformed RF data resulting in a signal-to-noise ratio (SNR) of 12 dB.

**Table 2.** Imaging pulse repetition frequency (PRF) and resulting inter-frame (IF) displacements.

| Simulations | PRF | IF Displacement Range (µm) |
|---|---|---|
| Blood flow | 15 kHz | 0–100 |
| | 2.5 kHz | 0–600 |
| Vessel wall | 25 Hz | 0.5–11.5 |
| | 50 Hz | 2.5–72 |
| **Rotating Disk Experiments** | **PRF** | **IF Displacement Range (µm)** |
| Blood flow scenario | 8 kHz | 0–23 |
| | 4 kHz | 0–46 |
| | 2 kHz | 0–91 |
| | 1 kHz | 0–182 |
| Vessel wall scenario | 8 kHz | 0–23 |
| | 4 kHz | 0–46 |

### 2.3.2. Carotid Vessel Wall Displacements

Simulations of the carotid artery vessel wall were performed using deformations patterns as obtained from a patient-specific finite element model (FEM) of the carotid artery at the bifurcation [40]. A transversal image plane, containing the internal and external carotid artery, was simulated. The amplitude of the scatterers representing the carotid artery were set at two, the surrounding tissue scatterers had an amplitude of one. Specular reflections were mimicked by positioning scatterers at the lumen-vessel wall and vessel wall-surrounding tissue transitions, which enhanced the realism of the RF data. Simulations of the vessel wall were performed in early systole, where peak inter-frame displacements are present and in late diastole, where low inter-frame displacements are present. A pre and post deformation frame were simulated for each time point, using a PRF of 50 Hz and 25 Hz (see Table 2) in systole and diastole, respectively. Band limited noise was added to the beamformed RF data resulting in an SNR of 30 dB.

### 2.4. Rotating Disk Experiments

A rotating disk experimental setup was built, providing a range of velocities at all possible beam-to-flow angles, or, a range of inter-frame displacements at full 360° range of directions. This type of setup has been used to study the performance of velocity estimation methods [29,43].

A cylindrical homogeneous phantom was created using a 15% polyvinyl alcohol (PVA) solution [39]. The solution was poured into a cylindrical mold (Øinner = 20 mm), containing a fixating-structure (see Figure 2) to attach the phantom onto the motor unit during the experiments, and subjected to three freeze-thaw cycles. Shrinkage of the phantom was present during assemblage, resulting in a final diameter of 18.2 mm. The phantom was attached to a motor unit (Closed Loop Step Motor, ARM69AC, Oriental Motor USA Corp., Torrance, CA, USA) and placed in a water tank (see Figure 2). It was rotated at a constant angular velocity, resulting in a maximum velocity at the outer edge of 18.2 cm/s. The phantom was positioned perpendicular to the transducer's scan plane at a depth of 3.4 cm. Plane wave data were acquired continuously using a PRF of 8 kHz for 4 s.

Two different approaches were used for processing the acquired data. One approach resembling flow velocity estimation and the second approach analogous to the processing performed for the vessel wall displacement simulations.

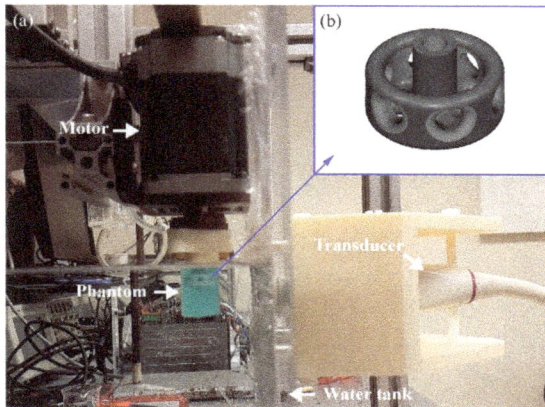

**Figure 2.** (a) Experimental setup rotating disk experiments. The cylindrical phantom is attached to the motor unit, using the fixating structure (**b**) and placed in a water tank. The transducer is positioned perpendicular to the phantom, at 3.4 cm from the center of the phantom.

### 2.4.1. Rotational Flow

After RF data beamforming and prior to velocity estimation, clutter filtering was performed. Although the phantom does not contain actual blood or tissue, including clutter filtering will give a more realistic picture of the performance of the methods for velocity estimation. A finite impulse response (FIR) filter with an order of 46 was used in combination with a temporal sliding window, not sacrificing frames for filter initialization. The −3 dB velocity cut-off point was equal to 0.068 times the Nyquist velocity. By down sampling the beamformed RF data in temporal direction, a range of PRFs (8, 4, 2 and 1 kHz) was studied. The resulting inter-frame displacements are summarized in Table 2. The same filter characteristics were used for all PRFs and estimated velocities were median averaged over 40 temporal frames.

### 2.4.2. Rotational Vessel Wall Displacement

Inter-frame displacements were estimated after RF data beamforming. Both PRFs of 8 and 4 kHz were studied, resembling inter-frame displacements present in early systole and late diastole of the vessel wall dynamics (see Table 2).

### 2.5. Beamforming Setup and Grid Definition

To be able to calculate the spacing for the PSF-shape-based beamforming grid, the PSF dimensions were determined for both the simulation and experimental setup (see Table 3). Field II was used to simulate the response of multiple point scatterers placed at spatial positions relevant for carotid artery imaging (Figure 3a). Simulated element data were beamformed on a high resolution grid and intensity curves for each individual scatterer were generated (Figure 3b). The axial ($PSF_{ax}$) and lateral ($PSF_{lat}$) size of the PSFs were determined based on the −6 dB values. The $PSF_{ratio}$ was determined according to $PSF_{ratio} = PSF_{lat}/PSF_{ax}$. The resulting ratios were averaged for all simulated scatterers. The PSF dimensions were determined experimentally using a wire phantom (60 μm diameter) placed in a water tank. Plane wave data were acquired of the wire phantom and similar processing was performed to determine the $PSF_{ratio}$.

Delay-and-sum beamforming was used to generate RF data for every plane wave transmission at pre-specified data sampling points [39]. The data were beamformed on the PSF-shape-based grid using the axial and lateral sampling distance as specified (Section 2.1.2) for the RF data ($d_{ax,RF}$ and $d_{lat,RF}$).

Consequently, the envelope data were obtained by demodulation and down sampling of the RF data in the axial direction. Furthermore, the element data were also beamformed on a conventionally spaced grid, with a lateral sampling distance equal to the pitch and 4 samples per wavelength in axial direction. Details on the grid spacing and beamforming parameters can be found in Table 3.

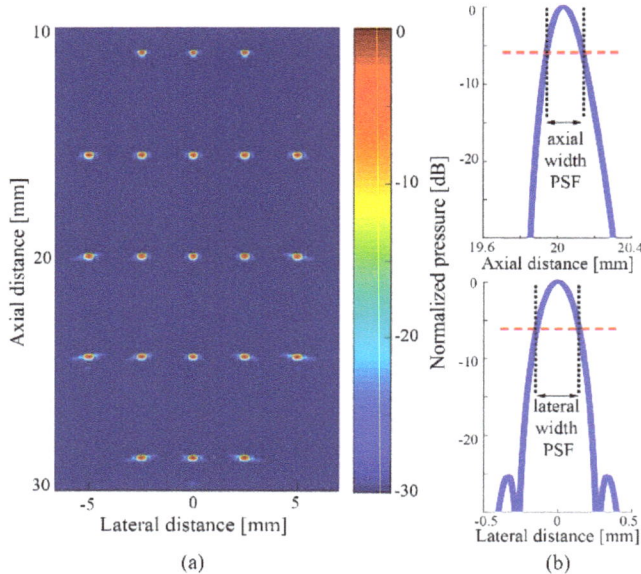

**Figure 3.** (**a**) Simulated PSFs at spatial positions relevant for carotid artery imaging. (**b**) Intensity curves in axial and lateral direction. The red lines indicate the −6 dB level.

**Table 3.** Beamforming parameters. PSF: point-spread-function.

| Grid Type | Parameter | Simulations | Experiments |
|---|---|---|---|
| PSF-shape-based grid | $PSF_{ratio}$ | 1.5 | 2.0 |
| | $d_{ax,RF}$ | 14.2 μm | 14.4 μm |
| | $d_{lat,RF}$ & $d_{lat,ENV}$ | 107 μm | 143 μm |
| | $d_{ax,ENV}$ | 71 μm | 72 μm |
| Conventional grid | $d_{ax}$ | 21.4 μm | 21.6 μm |
| | $d_{lat}$ | 198 μm | 198 μm |
| Both grids | Apodization window | Hamming | |
| | F-number | 0.875 | |

### 2.6. Multi-Step Speckle Tracking Settings

Inter-frame displacements were estimated using data beamformed on both the conventional grid and the PSF-shape-based grid. The size of the search windows was tuned to the maximum displacement expected in each situation. Spatial smoothing of the estimated displacements was performed using median filters. Inter-frame displacement estimates were multiplied by frame rate to calculate velocities.

Subsample resolution was resolved by interpolation of the CCF in both steps of the algorithm. Two types of interpolation were used: 1D parabolic fitting in each direction of the CCF independently, since it is one of the most often used interpolators in ultrasound applications as described in the Introduction, and 2D cubic interpolation, where a joint estimation of subsample displacement in both directions was obtained. The parabolic fitting is solved analytically using the commonly applied three-point parabola fitting [25,33]. 2D cubic interpolation was performed with a resulting effective upsampling factor of 10 and 1000 in the first and second step of the algorithm, respectively. For the second step, a fast in-house built implementation was used, where the high final resolution was reached iteratively by zooming in onto the peak of the CCF in multiple iteration steps. Given the CCF with its maximum at position (i,j), the new sampling points (ix,iy) are defined by ix = ixm*5$^{iter}$ + i and iy = iym*5$^{iter}$ + j, with matrices ixm and iym representing the direct neighbors of (i,j) according to ixm = [−1 0 1;−1 0 1;−1 0 1] and iym = [−1 −1 −1; 0 0 0; 1 1 1]. Ten iteration steps (iter = 1–10) were used, resulting in the effective upsampling factor of 1000. A complete overview of the settings used in the ST algorithm can be found in Table 4.

**Table 4.** Settings for 2-step speckle tracking algorithm.

| Parameter | Step | Blood Flow Simulations | Vessel Wall Simulations | Blood Flow Experiments | Vessel Wall Experiments |
|---|---|---|---|---|---|
| Template[1] | 1 | 0.84 × 2.20 | 0.93 × 0.77 | 0.9 × 1.6 | |
| - | 2 | 0.63 × 2.00 | 0.32 × 0.38 | 0.42 × 1.0 | |
| Search window[1] | 1 | 2.31 × 8.14 | 1.01 × 1.63 | 1.34 × 3.0 | |
| - | 2 | 0.74 × 2.85 | 0.43 × 1.24 | 0.54 × 2.14 | |
| 2D median filter[2] | 1 & 2 | 5 × 3 [3] | 5 × 3 [3] | 12 × 8 [3] | |
| - | - | 5 × 5 [4] | 5 × 5 [4] | 12 × 11 [4] | |

[1] Axial × lateral window size (mm); [2] Axial × lateral window size (samples); [3] Settings used for data on conventional grid; [4] Settings used for data on PSF-shape-based grid.

Summarized, four different combinations of beamforming grid and CCF interpolation type were studied in this work: (1) data beamformed on the conventional grid combined with 1D parabolic interpolation (conv_1Dpar) and (2) combined with 2D cubic interpolation (conv_2Dcub), (3) data beamformed on the PSF-shape-based grid combined with 1D parabolic interpolation (PSF_1Dpar) and (4) combined with 2D cubic interpolation (PSF_2Dcub).

The performance of the four approaches was studied in terms of the accuracy and precision of the velocity magnitude, velocity angle and the displacement estimates. The accuracy of the velocity magnitude and the inter-frame displacement estimates was assessed by calculating the absolute error percentage between the mean estimates and the ground truth as follows:

$$\text{Error (i)} = \frac{|EST_{mean}(i) - GT(i)|}{GT(i)} * 100\% \tag{1}$$

with $i$ all estimated samples within the region-of-interest (ROI), $EST_{mean}$ the mean of the estimates and $GT$ the ground truth value. $EST_{mean}$ was based on $n = 10$ realizations for the simulated parabolic flow, $n = 100$ ensemble averaged velocity estimates for the experimental rotational flow, and $n = 100$ inter-frame displacement estimates for the rotational displacement experiments. The error for the velocity angle estimates was calculated as the absolute error, i.e., error (i) = | $EST_{mean}(i) - GT(i)$ |.

The precision for the velocity magnitude and inter-frame displacement estimates was assessed by calculating the relative standard deviation:

$$\text{SD (i)} = \frac{\sigma_{EST}(i)}{GT_{max}} * 100\% \tag{2}$$

with $\sigma_{EST}$ the standard deviation of the same n estimates, and $GT_{max}$ the maximal ground truth value. The precision of the velocity angle estimates was calculated as the standard deviation ($\sigma_{EST}$). For the analysis of the experiments, the center of the disk (r < 1 mm) was excluded from the calculations to avoid velocities or displacements close to zero.

The performance for all methods for the vessel wall simulation was quantified by calculating the root-mean-squared error (RMSE) between estimated and ground truth axial and lateral displacements:

$$\text{RMSE} = \sqrt{\frac{1}{N} \sum_{i=1}^{N} (EST(i) - GT(i))^2} \tag{3}$$

with $N$ the number of estimated samples within the ROI, $EST(i)$ the estimated value for sample $i$, and $GT(i)$ the ground truth value for sample $i$.

## 3. Results

### 3.1. Simulations

#### 3.1.1. Parabolic Flow in Vessel Phantom

In Figure 4, the absolute bias and SD values are visualized as a function of the PRF for the parabolic flow simulations. The statistics were calculated while taking into account all simulated beam-to-flow angles. When the PRF increases, i.e., when the inter-frame displacements become smaller (see Table 2), the bias and SD for both the magnitude and angle increase. This effect is most dominant for the conventional grid, while the PSF-shape-based grid shows a more stable performance over the PRF range. Both PSF_1Dpar and PSF_2Dcub show high accuracy and precision, i.e., low bias and SD values, for the entire range of PRFs. For both methods, the median magnitude bias is below 10% and the angle bias remains below 0.55°.

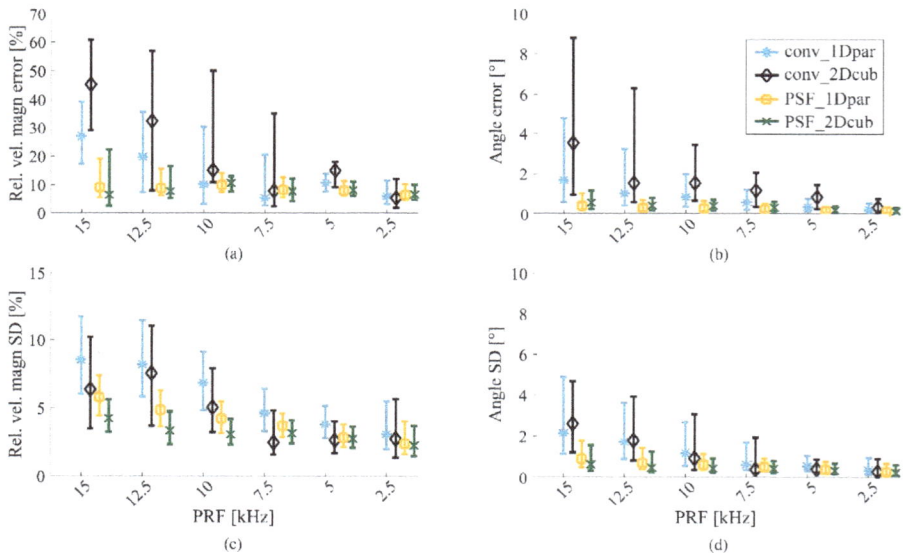

**Figure 4.** Median and interquartile ranges of the absolute bias (**a**) and standard deviation (SD) (**c**) for the velocity magnitude, and the absolute bias (**b**) and SD (**d**) for the velocity angle as a function of PRF. Statistics were calculated taking into account all beam-to-flow angles.

Figure 5 visualizes the dependency of the performance of the methods on beam-to-flow angle. The results were obtained using a PRF of 12.5 kHz. The accuracy and precision of the velocity angle estimates show a dependency on the beam-to-flow angle, whereas the magnitude estimate is less influenced by this. When the velocity field contains larger axial velocity components, i.e., for beam-to-flow angles different from 90°, the angle bias and SD values increase, especially for conv_1Dpar and conv_2Dcub. The PSF-shape-based grid combined with either 1D parabolic or 2D cubic interpolation shows a stable performance over the range of angles, with almost no increase of angle error and SD for angles other than 90°.

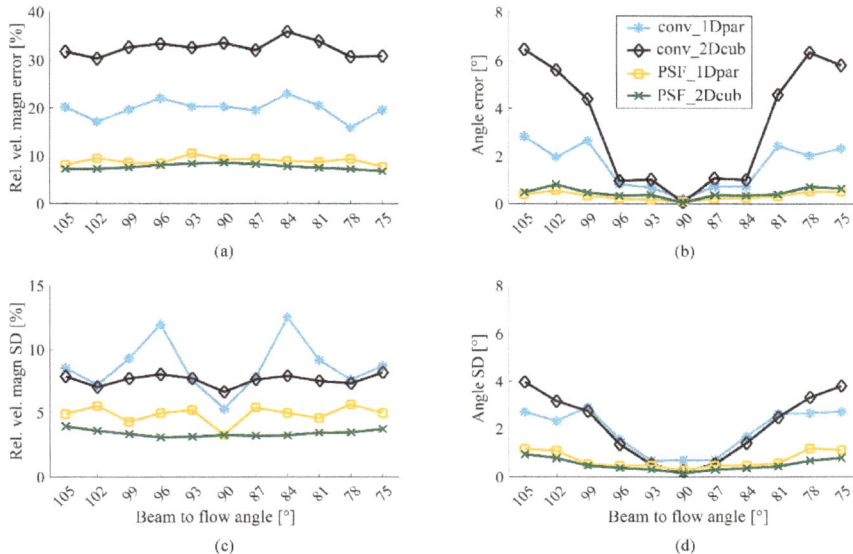

**Figure 5.** Median absolute bias and SD for the velocity magnitude (**a**, **c**) and angle (**b**, **d**), obtained using a PRF of 12.5 kHz. The performance of the four methods is visualized for a range of beam-to-flow angles.

Figures 4 and 5 show a competitive performance for PSF_1Dpar and PSF_2Dcub. Both methods show a stable performance for all PRFs, i.e., for 0 to 600 μm inter-frame displacements (see Table 2), and for all beam-to-flow angles. However, significantly lower median SD values for both velocity magnitude and angle can be appreciated for the PSF_2Dcub method (Wilcoxon signed-rank, $p < 0.05$). The median magnitude SD is at maximum 5.8% for PSF_1Dpar and 4.2% for PSF_2Dcub. Furthermore, the median angle SD is maximally 0.9° and 0.6° for PSF_1Dpar and PSF_2Dcub, respectively.

### 3.1.2. Carotid Vessel Wall Displacements

Figure 6 visualizes the estimated inter-frame displacements in the external and internal carotid vessel wall using the four different methods. Major performance difference can be seen for the lateral displacement estimation, with PSF_2Dcub showing the highest similarity with the ground truth. The RMSE values are visualized in Figure 7. The accuracy of the displacement estimates in the axial direction is higher than in the lateral direction, showing lower RMSE values for all four methods. This can also be appreciated from Figure 6. The largest performance differences between the four methods are observed for the lateral estimates. The PSF_2Dcub method shows lowest RMSEs for both the diastolic and systolic phase.

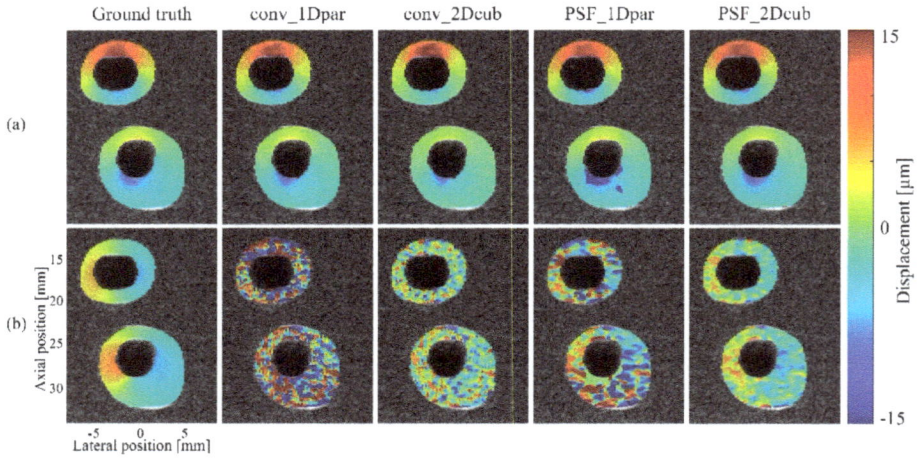

**Figure 6.** Axial (**a**) and lateral (**b**) inter-frame displacements for the carotid vessel phantom at diastole. The estimated displacements obtained using the four different methods are visualized together with the ground truth displacements derived from the finite element model (FEM) model.

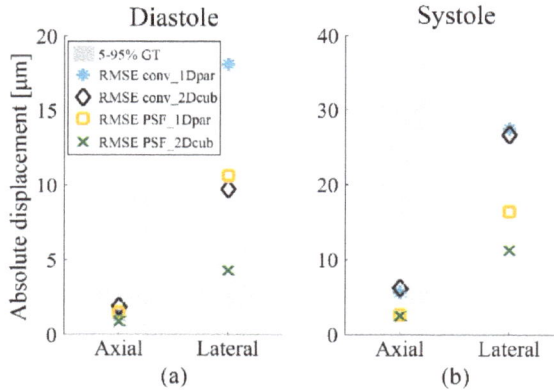

**Figure 7.** Root-mean-squared error (RMSE) values for the axial and lateral inter-frame displacement estimates obtained using the four different methods. Results are visualized for the diastolic (**a**) and systolic (**b**) phase. Absolute inter-frame ground truth (GT) 5–95% displacement ranges are also included in the figure. Please note the axial GT bar is cut off at 40 μm, while it ends at 72 μm.

## 3.2. Rotating Disk Experiments

### 3.2.1. Rotational Flow

Figure 8 shows the estimated and reference velocity magnitude and angle for the rotating disk experiment. Single ensemble-averaged velocity estimates for a PRF of 4 kHz are shown. Highest resemblance between velocity magnitude and ground truth is found for the PSF_2Dcub method. Here, largest deviations from the ground truth are observed in the center vertical slice of the disk, where velocities are mainly in lateral direction. This is a combined effect of the lower lateral resolution and clutter filtering. Due to the smaller axial velocity component in these regions, the estimates will be more influenced by the clutter filter [49,50]. In the bottom row of Figure 8, the angle estimation performance is visualized. Both the conv_2Dcub and PSF_2Dcub methods show good resemblance with the ground truth velocity angle.

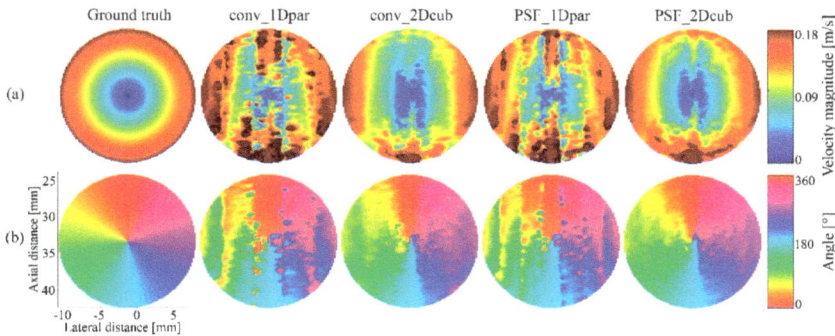

**Figure 8.** Estimated and ground truth velocity magnitude (**a**) and angle (**b**) obtained from the rotating disk phantom processed as rotational flow. The visualized results are obtained using a PRF of 4 kHz.

In Figure 9, the absolute bias and SD values are visualized as a function of the PRF. Similar effects can be seen as in the parabolic flow simulations: when inter-frame displacements become smaller, the bias and SD for both velocity magnitude and angle increase. This is most profound for the conv_1Dpar and PSF_1Dpar methods. The conv_2Dcub method shows a stable performance over the range of PRFs, although slightly higher bias and SD values are found as compared to the PSF_2Dcub method, especially for a PRF of 8 and 4 kHz. The PSF_2Dcub method shows a stable performance over the PRF-range, with low bias and SD values. Even for very small inter-frame displacements, the accuracy and precision of this method are excellent. A median velocity magnitude bias of maximally 10% and a median angle bias of 9.2° are found. The median SD for the velocity magnitude and angle are maximally 4.6% and 6.5°, respectively.

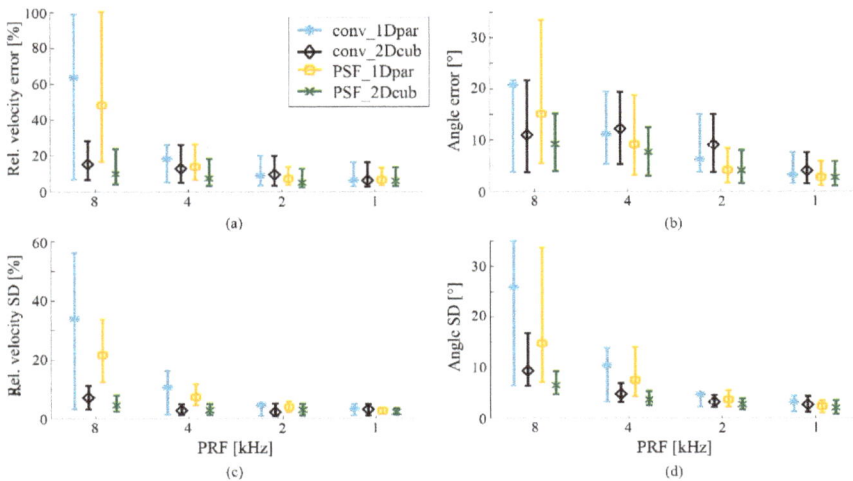

**Figure 9.** Median and interquartile ranges of the absolute bias and SD for the velocity magnitude (**a**, **c**) and angle (**b**, **d**) obtained from the rotating disk phantom processed as rotational flow. The performance of the methods is visualized as a function of the PRF.

### 3.2.2. Rotational Vessel Wall Displacements

In Figure 10, results for the rotating disk experiments are visualized when processed as vessel wall. Similar as in the vessel wall simulation study, the performance of the four methods is comparable for the axial displacement estimation. Furthermore, the axial accuracy is higher as the lateral displacement estimation accuracy. The lowest median lateral displacement error can be appreciated for the PSF_2Dcub method, although interquartile ranges of all methods show overlap. Furthermore, the precision of the lateral displacement estimation is highest for the PSF_2Dcub and conv_2Dcub methods. Overall, the PSF_2Dcub method shows the most stable performance with respect to the PRFs.

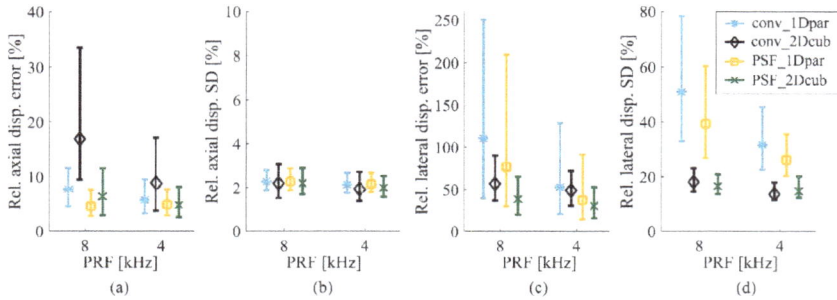

**Figure 10.** Median and interquartile ranges of the absolute bias and SD for the axial (**a**, **b**) and lateral (**c**, **d**) inter-frame displacements obtained from the rotating disk setup processed as rotational vessel wall displacements. The performance of the methods is visualized as function of the PRF, which corresponds to a range of inter-frame displacement (Table 2).

## 4. Discussion

In this paper, a new method has been proposed to determine the beamforming grid for ultrafast data based on the dimensions of the imaging system's PSF. As a result of this data sampling, the CCFs used in the ST algorithm will contain circular-shaped peaks. It was hypothesized that once combined with 2D cubic interpolation, this would increase the accuracy and precision of subsample displacement estimation and thereby enlarge the range of detectable displacements and velocities. Both simulations and experiments were conducted to characterize the performance of the method. A comparison was made with conventionally spaced beamformed data and the very commonly used 1D parabolic subsample displacement estimation method. The results confirmed our hypothesis: the PSF-shape-based beamforming grid combined with 2D cubic interpolation (PSF_2Dcub) showed the most accurate and stable performance with respect to the range of inter-frame displacements, both for the assessment of blood flow and vessel wall dynamics. Besides, the performance of the method was least affected by motion direction or beam-to-flow angle.

An important finding was that largest performance differences between the four methods were found for the lateral displacement component. This can be appreciated for the vessel wall (Figures 6, 7 and 10), but also for the flow results (Figure 8), in the upper and lower part of the disk, where velocities are predominantly in lateral direction. Furthermore, the accuracy and precision of the axial displacement estimates is consistently higher as compared to the lateral estimates, due to the higher resolution and presence of phase information. This also explains why the performance difference is less obvious in the axial direction and very clearly visible in the lateral direction. No phase information is available and the frequency content is lower in the lateral direction. Consequently, more benefit is gained from the proposed methodology.

Several functions can be used to interpolate the CCF for subsample displacement estimation. In this work, we used 1D parabolic interpolation, one of the most widely used functions, to compare to our proposed methodology. The results however, show that 1D parabolic is less optimal for subsample displacement estimation. Large bias, SD and RMSE values were found when 1D parabolic interpolation was used for CCF interpolation. More specifically, large bias and SD values on the lateral axis were shown when subsample displacements existed on the axial axis. This effect can be appreciated in Figure 8, where vertically stripped patterns are visible in the magnitude and angle estimates for the PSF_1Dpar en conv_1Dpar methods. These patterns originate from the inaccurate lateral displacement estimates in regions of subsample axial displacements, as also reported earlier [33]. The benefit of using 1D parabolic interpolation is the low computational cost, whereas 2D cubic interpolation increases computational time. Therefore, the trade-off between computational time and performance needs to be considered.

One limitation of this study is that a single 2D interpolator was investigated. Many other studies have investigated the use of other shapes and functions, such as cosine, parabolic or spline [25,33,35,51,52], most often in 1D. However, none of these interpolators have been studied before in the context of the proposed methodology, i.e., in combination with the freedom of choosing the beamforming grid and thereby influencing the CCF shape. Further investigation should be performed to compare different 2D interpolators in combination with the PSF-shape-based grid. 2D fitting of the CCF peak instead of interpolation is not considered in this work, due to the enormous increase of computational time, which makes its use less practical.

Another method to improve subsample displacement estimation is by interpolation of the echo signal. This method was often used in combination with conventionally acquired data and reduced the error by the up sample factor. However, the computational cost increased significantly [25,53]. The benefit of the use of ultrafast data is that instead of interpolating the echo signals, actual focusing of the signals can be performed in receive, as a post-processing step. This work shows that no excessive sampling of the RF data is required when using knowledge of the PSF dimensions to beamform the data combined with a suitable 2D subsample interpolation method.

The proposed methodology is based upon the statement that the CCF shape is primarily determined by the shape of the PSF. While it is one of the most important factors influencing the appearance of the CCF, it is known that other factors will also be of influence. In general, the performance of motion estimation methods will degrade as a result of signal decorrelation. Factors such as out-of-plane motion, presence of strong gradients within the displacement field and large deformations can hamper the match of the signal template within the search window, thereby influencing the appearance of the CCF. Smartly choosing data type and window sizes is recommended in those situations [25,54] and aligning and stretching methods are suggested to minimize signal decorrelation [55,56]. Although these and other factors will also influence the CCF shape, the proposed methodology was successful under the circumstances tested in this work.

To determine the spacing of the PSF-shape-based beamforming grid, a single average $PSF_{ratio}$ was used in this work. This was allowed since for carotid artery imaging, the PSF is nearly constant within the region of interest. However, for other applications, such as cardiac imaging, the size of the PSF will change with respect to the position in the imaging plane, amongst others due to the frequency dependent attenuation and location of the echo with respect to the ultrasonic beam [34]. These effects can be seen in the work of Tong et al. [57]. However, application of the PSF-shape-based grid is still possible, when using variable grid spacing tuned to the PSF shape locally. Nillesen et al. [58] showed first results using such a multi-zone, depth-dependent PSF-shape-based grid for cardiac motion imaging. The beamforming grid resolution was adjusted to the local PSF shape. This work also shows that the proposed beamforming strategy is not limited to linear array imaging. The PSF-shape-based beamforming grid can easily be used for data obtained with different transducer types, such as a phased array. The only requirement for using this method is the characterization of the imaging system by determining the (local) dimensions of the PSF.

In this work, the dimensions of the PSFs were determined using the response of single scatterers. When processing data acquired in vivo, a more sophistic approach could be favorable, which includes other factors influencing the shape of the PSF, such as attenuation. A representative estimate of the

PSF$_{ratio}$ could be obtained by using the spatial auto covariance method [59] to determine the speckle size in a B-mode image. Since speckle size is proportional to the dimensions of the PSF [60,61], this will provide the required information to determine the PSF-shape based beamforming grid.

In the present work, a large range of inter-frame displacements were studied by imaging with different PRFs. A stable and accurate performance of the proposed method was shown with respect to this large range of displacements. This shows the applicability of the proposed methodology for a broad field of research: it allows to study the dynamic behavior of multiple tissue types, such as the carotid artery, the heart and breast tissue, where very small displacements need to be captured, but also very large displacements and deformations can be present simultaneously.

In this work, the proposed beamforming strategy was combined with cross-correlation-based ST. However, this strategy is also applicable to other pattern matching functions, for example the sum of absolute squared differences. Furthermore, while this work presents a methodology in 2D, it could be easily adapted to 3D. The system's PSF needs to be characterized in 3 dimensions and interpolation of the 3D CCF in 3 directions should be performed simultaneously.

Zero-degree plane waves were used to acquire data in this work. The estimation of vessel wall displacements and blood flow velocities could be further improved by performing displacement compounding [40,49,62], a method introduced by Techavipoo et al. [63] and adapted for vascular strain estimation [64] and blood flow imaging [52]. Angled plane waves are used to derive the displacement or velocity vector, using solely axial displacement estimates. However, to improve the accuracy and precision of the estimates, often 2D ST is performed to acquire these axial displacements [62,64,65], meaning the accuracy of both the lateral and axial displacement estimates influence the performance of the compounding method. Therefore, displacement compounding would also strongly benefit from the methodology as presented in this work.

## 5. Conclusions

In this paper, a framework for robust displacement estimation in ultrafast ultrasound data was presented. It can be used as a protocolled way to beamform ultrafast data and obtain accurate estimates of the tissue displacements. The success of this new methodology is based on the following important concepts: the matching shape of the CCF peak and the interpolation function, the equal number of sample points in axial and lateral direction throughout the peak, and the use of 2D instead of 1D interpolation functions to acquire joint estimation of subsample displacements in both directions. The proposed method defines the beamforming grid based on the system's PSF to generate standardized circular-shaped CCFs which match with a 2D cubic interpolation function. The results presented in this work confirm the hypothesis; increased accuracy and precision were found for 2D motion estimation using the PSF-shape-based beamforming grid combined with 2D cubic subsample interpolation of the CCF. This was shown for a wide range of displacements or velocities present in carotid vessel wall and blood flow dynamics.

**Acknowledgments:** This research is supported by the Dutch Technology Foundation STW (NKG 12122), which is part of the Netherlands Organization for Scientific Research (NWO), and which is partly funded by the Ministry of Economic Affairs. The authors want to gratefully acknowledge the support of A. Nikolaev, MSc for his assistance in the rotating disk experiments.

**Author Contributions:** Anne E. C. M. Saris, Stein Fekkes, Maartje M. Nillesen, Hendrik H. G. Hansen and Chris L. de Korte conceived and designed the experiments/simulations; Anne E. C. M. Saris performed the simulations with support of Stein Fekkes, Anne E. C. M. Saris performed the experiments; Anne E. C. M. Saris analyzed the data with support of Stein Fekkes, Maartje M. Nillesen, Hendrik H. G. Hansen and Chris L. de Korte; Anne E. C. M. Saris is the main author of the paper; critical revisions were provided by all co-authors.

**Conflicts of Interest:** The authors declare no conflicts of interest.

# References

1. Evans, D.H.; Jensen, J.A.; Nielsen, M.B. Ultrasonic colour doppler imaging. *Interface Focus* **2011**, *1*, 490–502. [CrossRef] [PubMed]

2. D'Hooge, J.; Heimdal, A.; Jamal, F.; Kukulski, T.; Bijnens, B.; Rademakers, F.; Hatle, L.; Suetens, P.; Sutherland, G.R. Regional strain and strain rate measurements by cardiac ultrasound: Principles, implementation and limitations. *Eur. J. Echocardiogr.* **2000**, *1*, 154–170. [CrossRef] [PubMed]

3. Konofagou, E.E.; D'Hooge, J.; Ophir, J. Myocardial elastography—A feasibility study in vivo. *Ultrasound Med. Biol.* **2002**, *28*, 475–482. [CrossRef]

4. De Korte, C.L.; Pasterkamp, G.; van der Steen, A.F.W.; Woutman, H.A.; Bom, N. Characterization of plaque components using intravascular ultrasound elastography in human femoral and coronary arteries in vitro. *Circulation* **2000**, *102*, 617–623. [CrossRef] [PubMed]

5. Krouskop, T.A.; Wheeler, T.M.; Kallel, F.; Garra, B.S.; Hall, T. Elastic moduli of breast and prostate tissues under compression. *Ultrason. Imaging* **1998**, *20*, 260–274. [CrossRef] [PubMed]

6. Blessberger, H.; Binder, T. Non-invasive imaging: Two dimensional speckle tracking echocardiography: Basic principles. *Heart* **2010**, *96*, 716–722. [CrossRef] [PubMed]

7. Grubb, N.R.; Fleming, A.; Sutherland, G.R.; Fox, K.A. Skeletal muscle contraction in healthy volunteers: Assessment with doppler tissue imaging. *Radiology* **1995**, *194*, 837–842. [CrossRef] [PubMed]

8. Luo, J.; Li, R.X.; Konofagou, E.E. Pulse wave imaging of the human carotid artery: An in vivo feasibility study. *IEEE Trans. Ultrason. Ferroelectr. Freq. Control* **2012**, *59*, 174–181. [CrossRef] [PubMed]

9. Papadacci, C.; Pernot, M.; Couade, M.; Fink, M.; Tanter, M. High-contrast ultrafast imaging of the heart. *IEEE Trans. Ultrason. Ferroelectr. Freq. Control* **2014**, *61*, 288–301. [CrossRef] [PubMed]

10. Tanter, M.; Bercoff, J.; Sandrin, L.; Fink, M. Ultrafast compound imaging for 2-d motion vector estimation: Application to transient elastography. *IEEE Trans. Ultrason. Ferroelectr. Freq. Control* **2002**, *49*, 1363–1374. [CrossRef] [PubMed]

11. Ricci, P.; Maggini, E.; Mancuso, E.; Lodise, P.; Cantisani, V.; Catalano, C. Clinical application of breast elastography: State of the art. *Eur. J. Radiol.* **2014**, *83*, 429–437. [CrossRef] [PubMed]

12. Sadigh, G.; Carlos, R.C.; Neal, C.H.; Dwamena, B.A. Accuracy of quantitative ultrasound elastography for differentiation of malignant and benign breast abnormalities: A meta-analysis. *Breast Cancer Res. Treat.* **2012**, *134*, 923–931. [CrossRef] [PubMed]

13. Gamble, G.; Zorn, J.; Sanders, G.; MacMahon, S.; Sharpe, N. Estimation of arterial stiffness, compliance, and distensibility from m-mode ultrasound measurements of the common carotid artery. *Stroke* **1994**, *25*, 11–16. [CrossRef] [PubMed]

14. Schaar, J.A.; de Korte, C.L.; Mastik, F.; Strijder, C.; Pasterkamp, G.; Serruys, P.W.; van der Steen, A.F.W. Characterizing vulnerable plaque features by intravascular elastography. *Circulation* **2003**, *108*, 2636–2641. [CrossRef] [PubMed]

15. Hansen, H.H.; de Borst, G.J.; Bots, M.L.; Moll, F.L.; Pasterkamp, G.; de Korte, C.L. Compound ultrasound strain imaging for noninvasive detection of (fibro)atheromatous plaques: Histopathological validation in human carotid arteries. *JACC Cardiovasc. Imaging* **2016**, *9*, 1466–1467. [CrossRef] [PubMed]

16. Hansen, H.H.; de Borst, G.J.; Bots, M.L.; Moll, F.L.; Pasterkamp, G.; de Korte, C.L. Validation of noninvasive in vivo compound ultrasound strain imaging using histologic plaque vulnerability features. *Stroke* **2016**, *47*, 2770–2775. [CrossRef] [PubMed]

17. Klingelhofer, J. Ultrasonography of carotid stenosis. *Cerebrovasc. Dis.* **2013**, *35*, 1. [CrossRef]

18. Greene, E.R.; Eldridge, M.W.; Voyles, W.F.; Miranda, F.G.; Davis, J.G. Quantitative evaluation of atherosclerosis using doppler ultrasound. *IEEE Trans. Med. Imaging* **1982**, *1*, 68–78. [CrossRef] [PubMed]

19. Van der Worp, H.B.; Bonati, L.H.; Brown, M.M. Carotid stenosis. *N. Engl. J. Med.* **2013**, *369*, 2359. [PubMed]

20. Leitman, M.; Lysyansky, P.; Sidenko, S.; Shir, V.; Peleg, E.; Binenbaum, M.; Kaluski, E.; Krakover, R.; Vered, Z. Two-dimensional strain-a novel software for real-time quantitative echocardiographic assessment of myocardial function. *J. Am. Soc. Echocardiogr.* **2004**, *17*, 1021–1029. [CrossRef] [PubMed]

21. Trahey, G.E.; Allison, J.W.; von Ramm, O.T. Angle independent ultrasonic detection of blood flow. *IEEE Trans. Biomed. Eng.* **1987**, *34*, 965–967. [CrossRef] [PubMed]

22. Bohs, L.N.; Trahey, G.E. A novel method for angle independent ultrasonic imaging of blood flow and tissue motion. *IEEE Trans. Biomed. Eng.* **1991**, *38*, 280–286. [CrossRef] [PubMed]

23. Ophir, J.; Céspedes, I.; Ponnekanti, H.; Yazdi, Y.; Li, X. Elastography: A quantitative method for imaging the elasticity of biological tissues. *Ultrason. Imaging* **1991**, *13*, 111–134. [CrossRef] [PubMed]

24. Bohs, L.N.; Geiman, B.J.; Anderson, M.E.; Breit, S.M.; Trahey, G.E. Ensemble tracking for 2d vector velocity measurement: Experimental and initial clinical results. *IEEE Trans. Ultrason. Ferroelectr. Freq. Control* **1998**, *45*, 912–924. [CrossRef] [PubMed]

25. Lopata, R.G.; Nillesen, M.M.; Hansen, H.H.G.; Gerrits, I.H.; Thijssen, J.M.; de Korte, C.L. Performance evaluation of methods for two-dimensional displacement and strain estimation using ultrasound radio frequency data. *Ultrasound Med. Biol.* **2009**, *35*, 796–812. [CrossRef] [PubMed]

26. Bohs, L.N.; Geiman, B.J.; Anderson, M.E.; Gebhart, S.C.; Trahey, G.E. Speckle tracking for multi-dimensional flow estimation. *Ultrasonics* **2000**, *38*, 369–375. [CrossRef]

27. Shattuck, D.P.; Weinshenker, M.D.; Smith, S.W.; von Ramm, O.T. Explososcan: A parallel processing technique for high speed ultrasound imaging with linear phased arrays. *J. Acoust. Soc. Am.* **1984**, *75*, 1273–1282. [CrossRef] [PubMed]

28. Fadnes, S.; Nyrnes, S.A.; Torp, H.; Lovstakken, L. Shunt flow evaluation in congenital heart disease based on two-dimensional speckle tracking. *Ultrasound Med. Biol.* **2014**, *40*, 2379–2391. [CrossRef] [PubMed]

29. Fadnes, S.; Ekroll, I.K.; Nyrnes, S.A.; Torp, H.; Lovstakken, L. Robust angle-independent blood velocity estimation based on dual-angle plane wave imaging. *IEEE Trans. Ultrason. Ferroelectr. Freq. Control* **2015**, *62*, 1757–1767. [CrossRef] [PubMed]

30. Udesen, J.; Gran, F.; Hansen, K.L.; Jensen, J.A.; Thomsen, C.; Nielsen, M.B. High frame-rate blood vector velocity imaging using plane waves: Simulations and preliminary experiments. *IEEE Trans. Ultrason. Ferroelectr. Freq. Control* **2008**, *55*, 1729–1743. [CrossRef] [PubMed]

31. Jensen, J.A.; Nikolov, S.I.; Udesen, J.; Munk, P.; Hansen, K.L.; Pedersen, M.M.; Hansen, P.M.; Nielsen, M.B.; Oddershede, N.; Kortbek, J.; et al. Recent advances in blood flow vector velocity imaging. In Proceedings of the 2011 IEEE International, Ultrasonics Symposium (IUS), Orlando, FL, USA, 18–21 October 2011; pp. 262–271.

32. Swillens, A.; Segers, P.; Torp, H.; Lovstakken, L. Two-dimensional blood velocity estimation with ultrasound: Speckle tracking versus crossed-beam vector doppler based on flow simulations in a carotid bifurcation model. *IEEE Trans. Ultrason. Ferroelectr. Freq. Control* **2010**, *57*, 327–339. [CrossRef] [PubMed]

33. Zahiri Azar, R.; Goksel, O.; Salcudean, S.E. Sub-sample displacement estimation from digitized ultrasound rf signals using multi-dimensional polynomial fitting of the cross-correlation function. *IEEE Trans. Ultrason. Ferroelectr. Freq. Control* **2010**, *57*, 2403–2420. [CrossRef] [PubMed]

34. Céspedes, E.I.; Huang, Y.; Ophir, J.; Spratt, S. Methods for estimation of subsample time delays of digitized echo signals. *Ultrason. Imaging* **1995**, *17*, 142–171. [CrossRef] [PubMed]

35. Langeland, S.; D'Hooge, J.; Torp, H.; Bijnens, B.; Suetens, P. Comparison of time-domain displacement estimators for two-dimensional rf tracking. *Ultrasound Med. Biol.* **2003**, *29*, 1177–1186. [CrossRef]

36. Viola, F.; Walker, W.F. A spline-based algorithm for continuous time-delay estimation using sampled data. *IEEE Trans. Ultrason. Ferroelectr. Freq. Control* **2005**, *52*, 80–93. [CrossRef] [PubMed]

37. Luo, J.; Konofagou, E.E. Effects of various parameters on lateral displacement estimation in ultrasound elastography. *Ultrasound Med. Biol.* **2009**, *35*, 1352–1366. [CrossRef] [PubMed]

38. Swillens, A.; Segers, P.; Lovstakken, L. Two-dimensional flow imaging in the carotid bifurcation using a combined speckle tracking and phase-shift estimator: A study based on ultrasound simulations and in vivo analysis. *Ultrasound Med. Biol.* **2010**, *36*, 1722–1735. [CrossRef] [PubMed]

39. Hansen, H.H.; Saris, A.E.; Vaka, N.R.; Nillesen, M.M.; de Korte, C.L. Ultrafast vascular strain compounding using plane wave transmission. *J. Biomech.* **2014**, *47*, 815–823. [CrossRef] [PubMed]

40. Fekkes, S.; Swillens, A.E.; Hansen, H.H.; Saris, A.E.; Nillesen, M.M.; Iannaccone, F.; Segers, P.; de Korte, C.L. 2-d versus 3-d cross-correlation-based radial and circumferential strain estimation using multiplane 2-d ultrafast ultrasound in a 3-d atherosclerotic carotid artery model. *IEEE Trans. Ultrason. Ferroelectr. Freq. Control* **2016**, *63*, 1543–1553. [CrossRef] [PubMed]

41. Nayak, R.; Huntzicker, S.; Ohayon, J.; Carson, N.; Dogra, V.; Schifitto, G.; Doyley, M.M. Principal strain vascular elastography: Simulation and preliminary clinical evaluation. *Ultrasound Med. Biol.* **2017**, *43*, 682–699. [CrossRef] [PubMed]

42. Korukonda, S.; Nayak, R.; Carson, N.; Schifitto, G.; Dogra, V.; Doyley, M.M. Noninvasive vascular elastography using plane-wave and sparse-array imaging. *IEEE Trans. Ultrason. Ferroelectr. Freq. Control* **2013**, *60*, 332–342. [CrossRef] [PubMed]

43. Villagomez Hoyos, C.A.; Stuart, M.B.; Hansen, K.L.; Nielsen, M.B.; Jensen, J.A. Accurate angle estimator for high-frame-rate 2-d vector flow imaging. *IEEE Trans. Ultrason. Ferroelectr. Freq. Control* **2016**, *63*, 842–853. [CrossRef] [PubMed]

44. Saris, A.E.; Nillesen, M.M.; Fekkes, S.; Hansen, H.H.; de Korte, C.L. Robust blood velocity estimation using point-spread-function-based beamforming and multi-step speckle tracking. In Proceedings of the IEEE Ultrasonics Symposium, Taipei, Taiwan, 21–24 October 2015; pp. 1–4.

45. Lopata, R.G.P.; Nillesen, M.M.; Gerrits, I.H.; Thijssen, J.M.; Kapusta, L.; de Korte, C.L. In vivo 3d cardiac and skeletal muscle strain estimation. In Proceedings of the IEEE Ultrasonics, Vancouver, BC, Canada, 2–6 October 2006; pp. 744–747.

46. Shi, H.; Varghese, T. Two-dimensional multi-level strain estimation for discontinuous tissue. *Phys. Med. Biol.* **2007**, *52*, 389–401. [CrossRef] [PubMed]

47. Jensen, J.A.; Svendsen, N.B. Calculation of pressure fields from arbitrarily shaped, apodized, and excited ultrasound transducers. *IEEE Trans. Ultrason. Ferroelectr. Freq. Control* **1992**, *39*, 262–267. [CrossRef] [PubMed]

48. Jensen, J.A. Field: A program for simulating ultrasound systems. *Med. Biol. Eng. Comput.* **1996**, *34 Pt 1*, 351–353.

49. Saris, A.E.; Hansen, H.H.; Fekkes, S.; Nillesen, M.M.; Rutten, M.C.; de Korte, C.L. A comparison between compounding techniques using large beam-steered plane wave imaging for blood vector velocity imaging in a carotid artery model. *IEEE Trans. Ultrason. Ferroelectr. Freq. Control* **2016**, *63*, 1758–1771. [CrossRef] [PubMed]

50. Fadnes, S.; Bjaerum, S.; Torp, H.; Lovstakken, L. Clutter filtering influence on blood velocity estimation using speckle tracking. *IEEE Trans. Ultrason. Ferroelectr. Freq. Control* **2015**, *62*, 2079–2091. [CrossRef] [PubMed]

51. Lai, X.; Torp, H. Interpolation methods for time-delay estimation using cross-correlation method for blood velocity measurement. *IEEE Trans. Ultrason. Ferroelectr. Freq. Control* **1999**, *46*, 277–290. [PubMed]

52. Viola, F.; Coe, R.L.; Owen, K.; Guenther, D.A.; Walker, W.F. Multi-dimensional spline-based estimator (muse) for motion estimation: Algorithm development and initial results. *Ann. Biomed. Eng.* **2008**, *36*, 1942–1960. [CrossRef] [PubMed]

53. Konofagou, E.; Ophir, J. A new elastographic method for estimation and imaging of lateral displacements, lateral strains, corrected axial strains and poisson's ratios in tissues. *Ultrasound Med. Biol.* **1998**, *24*, 1183–1199. [CrossRef]

54. Keane, R.D.; Adrian, R.J. Theory of cross-correlation analysis of piv images. *Appl. Sci. Res.* **1992**, *49*, 191–215. [CrossRef]

55. Lopata, R.G.P.; Hansen, H.H.G.; Nillesen, M.M.; Thijssen, J.M.; Kapusta, L.; de Korte, C.L. Methodical study on the estimation of strain in shearing and rotating structures using radio frequency ultrasound based on 1-d and 2-d strain estimation techniques. *IEEE Trans. Ultrason. Ferroelectr. Freq. Control* **2010**, *57*, 855–865. [CrossRef] [PubMed]

56. Alam, S.K.; Ophir, J. Reduction of signal decorrelation from mechanical compression of tissues by temporal stretching: Applications to elastography. *Ultrasound Med. Biol.* **1997**, *23*, 95–105. [CrossRef]

57. Tong, L.; Gao, H.; Choi, H.F.; D'hooge, J. Comparison of conventional parallel beamforming with plane wave and diverging wave imaging for cardiac applications: A simulation study. *IEEE Trans. Ultrason. Ferroelectr. Freq. Control* **2012**, *59*, 1654–1663. [CrossRef] [PubMed]

58. Nillesen, M.M.; Saris, A.E.C.M.; Hansen, H.H.; Fekkes, S.; van Slochteren, F.J.; Bovendeerd, P.H.M.; De Korte, C.L. Cardiac motion estimation using ultrafast ultrasound imaging tested in a finite element model of cardiac mechanics. In *Functional Imaging and Modeling of the Heart*; Springer: Cham, Germany, 2015; pp. 207–214.

59. Wagner, R.F.; Smith, S.W.; Sandrik, J.M.; Lopez, H. Statistics of speckle in ultrasound b-scans. *IEEE Trans. Sonics Ultrason.* **1983**, *30*, 156–163. [CrossRef]

60. Wagner, R.F.; Insana, M.F.; Smith, S.W. Fundamental correlation lengths of coherent speckle in medical ultrasonic images. *IEEE Trans. Ultrason. Ferroelectr. Freq. Control* **1988**, *35*, 34–44. [CrossRef] [PubMed]

61. Thijssen, J.M.; Oosterveld, B.J. Texture in tissue echograms. Speckle or information? *J. Ultrasound Med.* **1990**, *9*, 215–229. [CrossRef] [PubMed]

62. He, Q.; Tong, L.; Huang, L.; Liu, J.; Chen, Y.; Luo, J. Performance optimization of lateral displacement estimation with spatial angular compounding. *Ultrasonics* **2017**, *73*, 9–21. [CrossRef] [PubMed]

63. Techavipoo, U.; Chen, Q.; Varghese, T.; Zagzebski, J.A. Estimation of displacement vectors and strain tensors in elastography using angular insonifications. *IEEE Trans. Med. Imaging* **2004**, *23*, 1479–1489. [CrossRef] [PubMed]

64. Hansen, H.H.; Lopata, R.G.; Idzenga, T.; de Korte, C.L. Full 2d displacement vector and strain tensor estimation for superficial tissue using beam-steered ultrasound imaging. *Phys. Med. Biol.* **2010**, *55*, 3201–3218. [CrossRef] [PubMed]

65. Xu, H.; Varghese, T. Normal and shear strain imaging using 2d deformation tracking on beam steered linear array datasets. *Med. Phys.* **2013**, *40*, 012902. [CrossRef] [PubMed]

*applied*
*sciences*

MDPI

*Article*

# Quasi-Static Elastography and Ultrasound Plane-Wave Imaging: The Effect of Beam-Forming Strategies on the Accuracy of Displacement Estimations

Gijs A.G.M. Hendriks [1,*], Chuan Chen [1], Hendrik H.G. Hansen [1] and Chris L. de Korte [1,2]

[1] Medical UltraSound Imaging Center, Department of Radiology and Nuclear Medicine, Radboud University Medical Center, P.O. Box 9101, 6500 HB Nijmegen, The Netherlands; chuan.chen@radboudumc.nl (C.C.); rik.hansen@radboudumc.nl (H.H.G.H.); chris.dekorte@radboudumc.nl (C.L.d.K.)

[2] Physics of Fluids Group, MIRA, University of Twente, P.O. Box 217, 7500 AE Enschede, The Netherlands

* Correspondence: gijs.hendriks@radboudumc.nl; Tel.: +31-24-365-1503

Received: 22 December 2017; Accepted: 17 February 2018; Published: 26 February 2018

**Abstract:** Quasi-static elastography is an ultrasound method which is widely used to assess displacements and strain in tissue by correlating ultrasound data at different levels of deformation. Ultrafast plane-wave imaging allows us to obtain ultrasound data at frame rates over 10 kHz, permitting the quantification and visualization of fast deformations. Currently, mainly three beam-forming strategies are used to reconstruct radio frequency (RF) data from plane-wave acquisitions: delay-and-sum (DaS), and Lu's-fk and Stolt's-fk operating in the temporal-spatial and Fourier spaces, respectively. However, the effect of these strategies on elastography is unknown. This study investigates the effect of these beam-forming strategies on the accuracy of displacement estimation in four transducers (L7-4, 12L4VF, L12-5, MS250) for various reconstruction line densities and apodization/filtering settings. A method was developed to assess the accuracy experimentally using displacement gradients obtained in a rotating phantom. A line density with multiple lines per pitch resulted in increased accuracy compared to one line per pitch for all transducers and strategies. The impact on displacement accuracy of apodization/filtering varied per transducer. Overall, Lu's-fk beam-forming resulted in the most accurate displacement estimates. Although DaS in some cases provided similar results, Lu's-fk is more computationally efficient, leading to the conclusion that Lu's-fk is most optimal for plane wave ultrasound-based elastography.

**Keywords:** quasi-static; elastography; ultrasound; beam-forming; delay-and-sum; Stolt's; Lu's; displacements; apodization; lateral displacement; axial displacement

---

## 1. Introduction

In 1991, Ophir et al. [1] proposed a new ultrasound technique, quasi-static elastography, to visualize and quantify deformation of tissue to thus detect abnormalities (e.g., through the relatively stiffer breast and prostate tumors). In this technique, ultrasound radio frequency (RF) data are obtained prior to and after deformation of the target tissue. Deformation can be induced externally by the transducer or a vibrator, or internally by the heart or respiration. Several methods have been developed to estimate displacement or strain maps based on the acquired RF data [2–6]. Conventionally, a template windows containing post-deformation RF-data is cross-correlated with a search window containing pre-deformation data. The position of the cross-correlation peak indicates the displacement of the template. To cope with relatively large displacements, coarse-to-fine cross-correlation methods were developed [2]. In coarse-to-fine cross-correlation, multi-iterative cross-correlation is executed starting with relatively large windows analyzing the envelope signal and subsequently using finer windows of RF-signal in each subsequent iteration with the estimated displacements of the previous iteration as an offset. Finally, a least squares strain estimator can be used to reconstruct a strain map [7].

The majority of current commercial ultrasound systems collect ultrasound RF data line-by-line using focused ultrasound transmit-and-receive sequences. Back in 1979, plane wave acquisitions were proposed as alternative to line-by-line sequences [8,9]. This type of acquisition has been getting more attention since the mid-2000s when the computational power and data transfer rates had sufficiently increased to enable the processing of the data generated by plane-wave transmissions. In plane-wave imaging, the full transducer array is used to transmit one unfocussed plane-wave and receive the reflected signals (Figure 1). In case of a 192-element array transducer, frame-rates can be increased by a factor of 192 compared to line-by-line scans. Consequently, ultrasound data can be collected with frame-rates over 10 kHz (theoretically limited by the two-way propagation of the ultrasound signal) and so fast deformations can also be quantified and visualized using quasi-static elastography. In the absence of focusing in transmit, plane-wave imaging requires software beam-forming to reconstruct RF-data from the obtained element data. In conventional ultrasound scanners, beam-forming is often applied in hardware. In this study, three software beam-forming strategies and their effect on displacement estimation were exploited: delay-and-sum (DaS) [10] in the spatial-temporal domain originating from conventional focused ultrasound, and Lu's-fk [11,12] and Stolt's-fk [13] beam-forming methods, both operating in the Fourier domain which are specifically developed for plane-wave acquisitions.

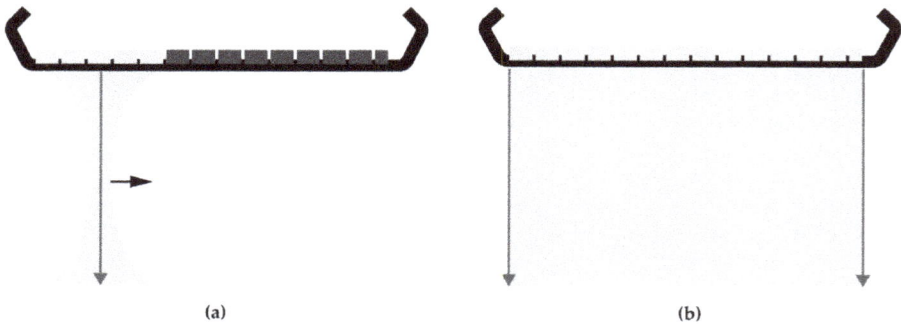

**(a)**                    **(b)**

**Figure 1.** Illustrations of: (**a**) a focused acquisition series in which the aperture (active elements in light gray) is shifted to acquire data line-by-line by transmitting a focused ultrasound beam and receiving the reflected signal for each aperture position; (**b**) a plane-wave acquisition, where an unfocused ultrasound is transmitted and reflected signals are received by the full transducer aperture.

DaS is a beam-forming method based on delaying the element signal by the expected time-of-flight (ToF) and summation of the element data points (Figure 2). For a certain point to be reconstructed, the ToF is calculated as the time of the plane-wave propagating to that point and the reflected or back-scattered signal to be received by the elements. The ToF is used to identify and sum the signals originating from each point in the element data. In dynamic focusing in receive, the F-number (F) dictates the number of elements (aperture) used for beam-forming:

$$F = \frac{d}{l_{app}} \tag{1}$$

where $d$ and $l_{app}$ are the depth of the reconstruction point and aperture width, respectively. Furthermore, apodization can be applied to weigh the signals (e.g., by Hamm or Hann function) to reconstruct each point and so expected angle sensitivity and back-scatter signal intensities can be incorporated. Apodization is often applied to increase contrast in B-mode images.

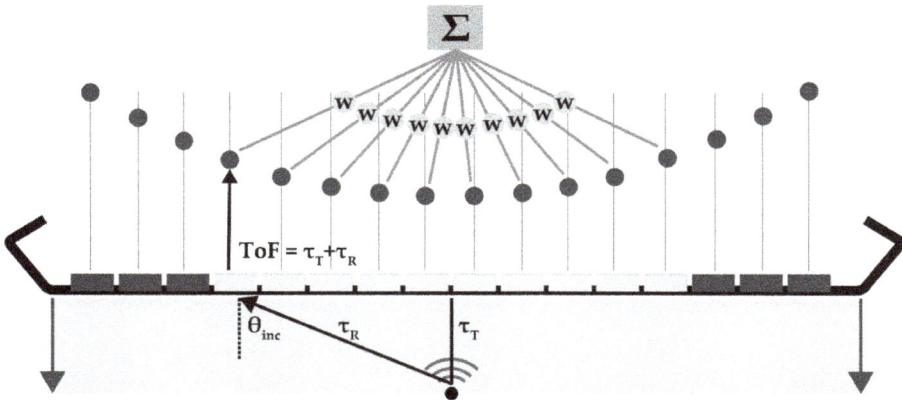

**Figure 2.** Schematic overview of delay-and-sum (DaS) beam-forming: for each reconstruction point, the time-of-flight (ToF) can be calculated by the transmit ($\tau_T$) and receive time towards each element ($\tau_R$). These ToF values are used to delay and sum the element data; if required the data can be weighted (w) by an apodization function. $\theta_{inc}$ is the incidence angle of the signal.

Another approach is beam-forming in the Fourier domain by Lu's-fk or Stolt's-fk which are computationally less expensive and can run around 25 times faster compared to DaS [13]. In both methods, 2-D element data in the spatial-temporal domain are transformed to the fk-space. Next, a migration method is applied to transform the data into the complete k-space and finally the migrated data are transformed back into the spatial-temporal domain (Figure 3). A migration method of plane-wave ultrasound was first carried out by Lu et al. from the analysis of limited diffraction beams [11,12]. Lu et al. introduced the mapping rule throughout the transmission–scattering–receiving process. More recently, Garcia et al. proposed another method by adapting Stolt's-fk seismic wave migration to the plane wave ultrasound scenario [13]. In this modification, the plane wave's backscattering is fitted with waves spontaneously radiating from sources. Comparing Stolt's-fk and Lu's-fk, the two methods share a similar spectrum content at small angles, while their spectra differ more significantly in the mapping of large angles.

Similar to apodization in DaS, angular weighting can be applied in Fourier domain beam-forming to increase contrast in the final B-mode image [14]. In this method, developed in our group, the Fourier spectrum after the migration (k-space) is multiplied with a template which is designed such that reflected waves originating from directions closely aligned with the plane-wave steering angle pass unaltered, whereas waves from wave directions deviating from the plane-wave propagation direction are attenuated according to a Hann function (Figure 3). This template can be applied for both migration methods (Lu's-fk and Stolt's-fk).

Since the beam-forming in plane-wave imaging is executed using software, the design of the beam-forming reconstruction grid (USGrid) is more flexible compared to conventional focused line-by-line scanners in which beam-forming is executed by hardware. Consequently, more lines per pitch can be reconstructed, which might result in more accurate displacement estimates. For focused imaging, it has already been shown that more accurate lateral displacement estimates can be obtained when increasing the RF line density by interpolation [15,16].

Plane-wave imaging and displacement and strain imaging have been combined in multiple studies. Besides a paper from our group [17] which investigated the effect of line density on displacement estimation for DaS, to our knowledge, the effects of different beam-forming strategies and line densities on displacement and strain estimation have not been explored. Therefore, the aim of this study was to investigate the effect of beam-forming strategies (DaS, Lu's-fk and Stolt's-fk) for plane-wave imaging on the accuracy of displacement estimates. Furthermore, the effect of the line density, apodization (in DaS) and angular weighting (in Lu's-fk and Stolt's-fk) were evaluated, and four different transducers were used with central frequencies varying between 5 and 21 MHz representing the whole range of clinically used linear array transducers. In this study, we evaluated the performance of displacement and strain estimation by using a rotating phantom. The advantage of a rotating phantom is that a displacement field is induced with axial and lateral displacements varying separately in magnitude instead of having pure axial or lateral displacements with same magnitude as obtained by linear translation of a phantom. Furthermore, the gradients of the axial and lateral displacement field were used to evaluate the accuracy of the displacement estimations because these gradients should be zero (see Materials and Methods section) without requiring exact knowledge of the reference displacement field. These gradients can also be considered as strains since strains are calculated the same way and are equal to their gradient.

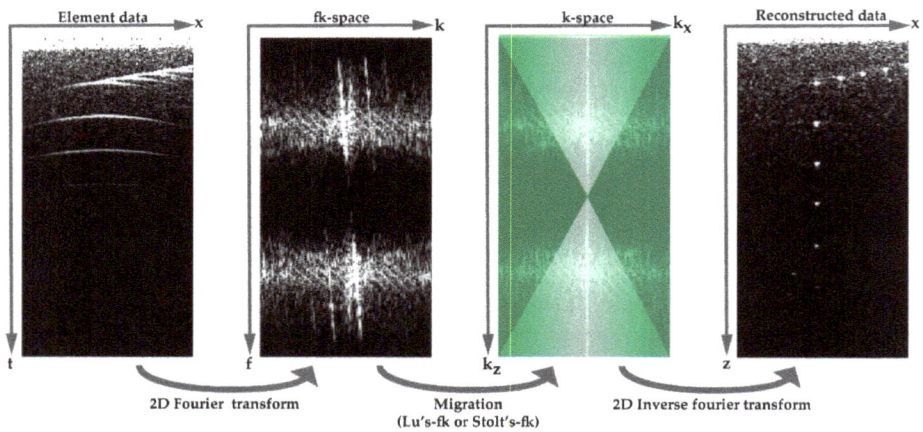

**Figure 3.** Overview of beam-forming in the fk-space: acquired element data is transformed to the fk-space using 2-D Fourier transform, and Lu's-fk or Stolt's-fk migration is applied to convert the data into the k-space. If required, this spectrum can be multiplied by a template (green overlay) to filter the data. Finally, the reconstructed data can be obtained by the 2-D inverse Fourier transform.

## 2. Materials and Methods

To create a block phantom, gelatin (10% by weight; VMR International, Leuven, Belgium) was dissolved in demineralized water and heated to 90 °C and cooled to 35 °C while continuously stirring using a magnetic stirrer. During the cooling process, silica particles (2% by weight; silica gel 60, 0.015–0.040 mm; Merck KGaA, Darmstadt, Germany) were added, which act as scatterers. Next, the mixed solution was poured in an open container (200 × 100 × 100 mm) leaving a 20 mm space under the top surface. Finally, the phantom was placed in a fridge to consolidate for 24 h.

The phantom, including the container, was positioned on top of a seesaw to enable rotation of the phantom. One of the four different linear array transducers utilized in this study (Table 1) was positioned above the phantom top surface such that the transducer footprint was unable to touch the phantom after rotation at the maximum angle of 10 degrees. Water was poured on the top surface of the phantom to fill the remaining 20 mm space and to ensure acoustic coupling between the transducer and phantom. The transducer was connected to a Verasonics V1 system (MS250 to Verasonics Vantage) to enable 0° plane-wave transmissions and element data collection using the center 128 elements of the transducer (MS250: all 256 elements). Element data were recorded prior to and after rotation (~2°) of the phantom. The experimental setup is also visualized in Figure 4.

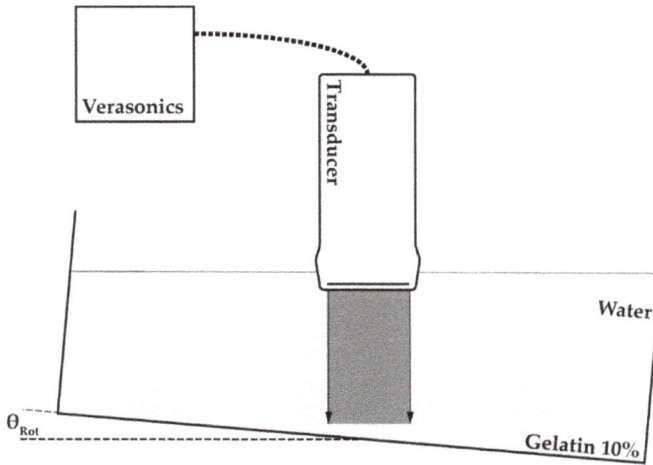

**Figure 4.** Experimental setup: transducer (Table 1) was connected to a Verasonics (V1 or Vantage) research ultrasound machine; element data were acquired prior to and after rotation ($\theta_{Rot}$) of a container in which a gelatin phantom was positioned (light grey) and water was poured on top of the phantom to ensure ultrasonic coupling.

**Table 1.** Transducers used in this study with their bandwidth, central frequency ($f_c$), and pitch.

| Transducer | Bandwidth | $f_c$ | Pitch | Manufacturer |
|:---:|:---:|:---:|:---:|:---:|
| L7-4 | 4–7 MHz | 5.0 MHz | 298 μm | ATL [1] |
| 12L4VF | 4–12 MHz | 8.2 MHz | 266 μm | Siemens [2] |
| L12-5 | 5–12 MHz | 9.0 MHz | 198 μm | ATL [1] |
| MS250 | 13–24 MHz | 21 MHz | 88 μm | VisualSonics [3] |

[1] ATL, Bothell, WA, USA; [2] Siemens Healthineers, Issaquah, WA, USA; [3] FUJIFILM VisualSonics Inc., Toronto, ON, Canada.

Element data were beam-formed using three different strategies: (1) Lu's-fk [11,12], Stolt's-fk [13], and DaS [10]; (2) with and without angular weighting (Lu's-fk; Stolts-fk) [14] or apodization by Hamm function (DaS); and (3) a beam-forming ultrasound grid (USGrid) with 8 samples per wave length and 1, 2, 3 and 4 lines per pitch. These strategies were implemented in Matlab (2016a; Mathworks Inc., Natick, MA, USA). Zero-padding was applied in the k-space to achieve higher line-densities for the fk-based strategies. Beam-forming settings were empirically determined: plane-wave element data were acquired in a multi-purpose phantom and beam-forming settings were tuned resulting in optimal contrast and resolution after beam-forming (see Appendix A for more details). For all transducers, the F-number (DaS) and angle weighting range (Lu's-fk; Stolts-fk) were set to 0.875 and ±20°, respectively.

Next, displacements were estimated by two step coarse-to-fine normalized cross correlation of pre- and post-rotation beam-formed ultrasound envelope and RF data in the first and second step respectively [18,19]. The cross-correlation peak was interpolated (2-D spline) after the final iteration to estimate sub-sample displacements. After each iteration, displacements were median filtered to remove outliers. Template and search windows, and other settings can be found in Table 2.

**Table 2.** Settings of the displacement estimation algorithm and displacement grid (DispGrid).

| Transducer | Step | Template Window [1] | Search Window [1] | Filter Size [2] | DispGrid [1] |
|:---:|:---:|:---:|:---:|:---:|:---:|
| L7-4 | 1 | 1.22 × 1.49 | 4.82 × 5.07 | 5 × 5 | 0.298 × 0.298 |
| - | 2 | 0.62 × 0.90 | 0.89 × 1.49 | 3 × 3 | 0.298 × 0.298 |
| 12L4VF | 1 | 0.68 × 1.33 | 2.68 × 4.52 | 5 × 5 | 0.266 × 0.266 |
| - | 2 | 0.35 ×0.80 | 0.49 × 1.33 | 3 × 3 | 0.266 × 0.266 |
| L12-5 | 1 | 0.68 × 0.99 | 2.68 × 3.37 | 5 × 5 | 0.198 × 0.198 |
| - | 2 | 0.35 × 0.60 | 0.49 × 0.99 | 3 × 3 | 0.198 × 0.198 |
| MS250 | 1 | 0.29 × 0.44 | 1.15 × 1.50 | 5 × 5 | 0.088 × 0.088 |
| - | 2 | 0.15 × 0.27 | 0.21 × 0.44 | 3 ×3 | 0.088 × 0.088 |

[1] Axial × lateral window size or DispGrid resolution in mm; [2] # samples and # lines in DispGrid which is independent of # lines per pitch.

The axial and lateral displacements ($u_z$ and $u_x$) after rotation can be described as:

$$u_x = (\cos(\theta) - 1)(x - x_0) - \sin(\theta)(z - z_0) \tag{2}$$

$$u_z = \sin(\theta)(x - x_0) + (\cos(\theta) - 1)(z - z_0) \tag{3}$$

where $\theta$ and $(x_0, z_0)$ are the rotation angle and center coordinates, respectively. The gradient of $u_x$ and $u_z$ in the lateral ($x$) and the axial ($z$) directions can respectively be described as:

$$s_{xx} = \frac{du_x}{dx} = \cos(\theta) - 1 \tag{4}$$

$$s_{zz} = \frac{du_z}{dz} = \cos(\theta) - 1 \tag{5}$$

In this experiment, angle $\theta$ can be considered small (~2°) and so it can be assumed that:

$$\cos(\theta) \approx 1 - \frac{\theta^2}{2} \approx 1 \tag{6}$$

Consequently, $s_{xx}$ and $s_{zz}$ in (Equations (4) and (5)) approximate zero, independently of the rotation angle and center. As these gradients (strains) yield 0 and the exact rotational angle and center were unknown in this experiment, we adopted the root-mean squared error (RMSE) of the gradients as a measure of the accuracy of the displacement estimates. The gradients of the displacements were calculated using a one-dimensional three-point least-squares strain estimator [7]. Gradients were only calculated and evaluated within a field-of-view (FoV) measuring –10 to 10 mm laterally and 17.5 to 52.5 mm axially for all transducers. However, the FoV measurement of the MS250 transducer was –8 to 8 mm laterally, since the FoV was limited by the transducer footprint. The top 17.5-mm axial depth was measured through water and was therefore neglected for all transducers. All acquisition and processing steps and intermediate results are summarized in Figure 5.

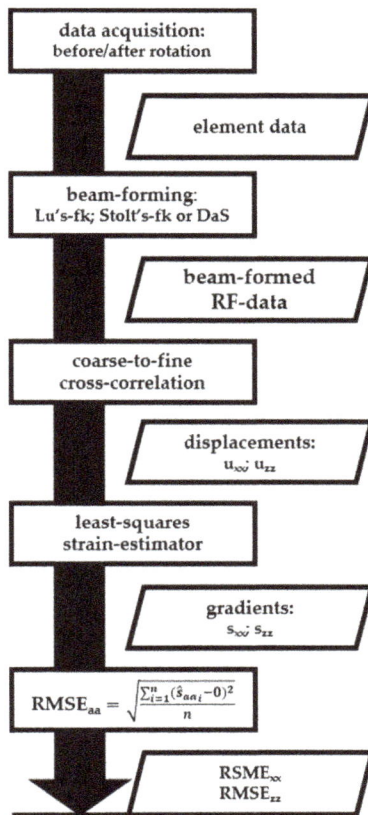

**Figure 5.** Summary of the processing flow described in the Material and Methods section. This work flow was repeated for every transducer (Table 1). Rectangles represent processing steps (acquisitions or calculations) and parallelograms represent (intermediate) results.

## 3. Results

Figure 6 provides an overview of the estimated axial and lateral displacement and strain fields for all probes using Lu's-fk beam-forming and a line density of two lines per pitch. As expected from Equations (2) to (6), the axial displacement fields (Figure 6a–d) were constant in the axial direction and so the gradient of the axial displacements in axial directions ($s_{zz}$) were approximately 0% (see Figure 6f–i). For the 12L4VF transducer the results revealed some artifacts in the left top corner (Figure 6b,g). These artifacts were visible for all beam-forming strategies and line densities. The MS250 transducer showed more outliers at lower depths (30–52.5 mm) which were probably caused by attenuation of the ultrasound signal at relatively large depths for this frequency.

The lateral displacement (Figure 6k–n) and strain fields (Figure 6p–s) resulted in similar observation as the axial results: constant displacement values in lateral direction and so approximately 0% gradient of lateral displacements in that direction ($s_{xx}$). Similar artefacts were also visible for the 12L4VF and MS250 transducers since the axial and lateral displacements were estimated using two-dimensional cross-correlations, which implies these estimations were coupled. Compared to the axial displacement and strain fields (Figure 6a–j), the lateral fields (Figure 6k–t) were noisier for all transducers, beam-forming strategies and line densities. In the case of two lines per pitch and Lu's-fk beam-forming as shown in Figure 6, $s_{zz}$ varied between $\pm 0.5\%$, whereas $s_{xx}$ varied between $\pm 1\%$.

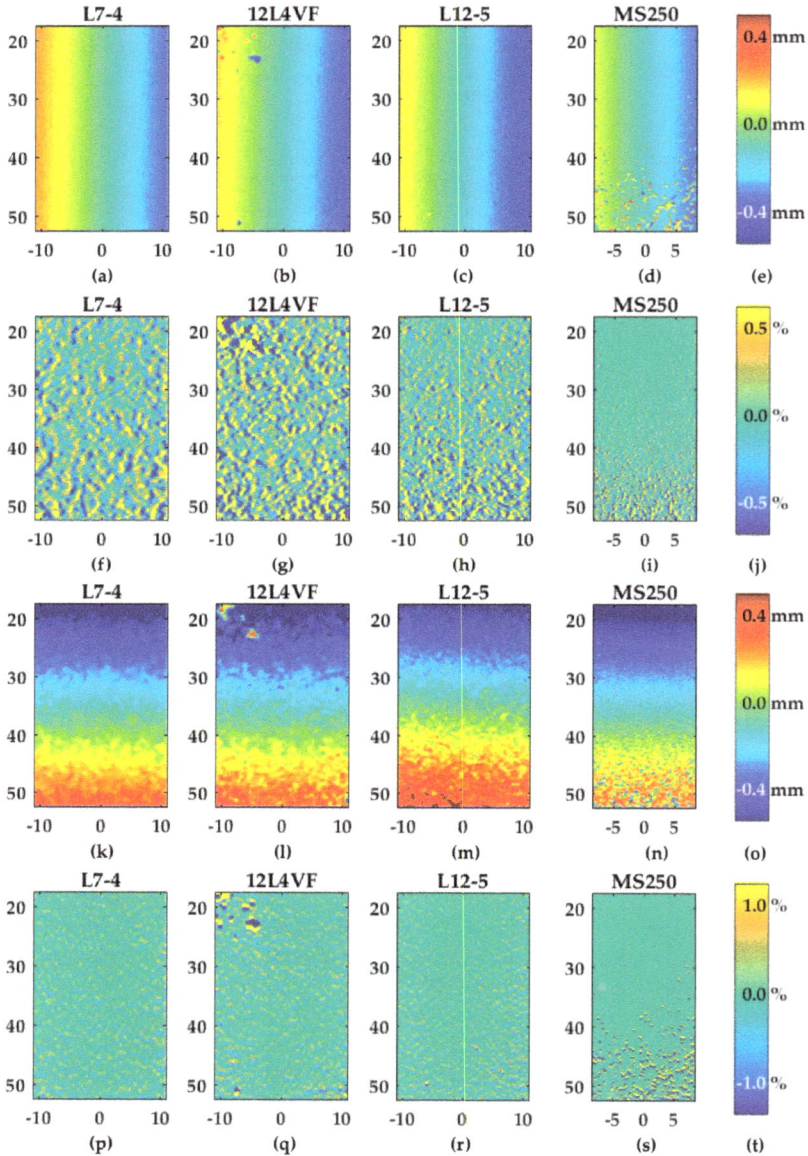

**Figure 6.** Overview of displacement and strain fields after rotation using Lu's-fk and two lines per pitch as a beam-forming strategy and for line density, respectively: (**a–c**) axial displacements; (**f–n**) axial strains ($s_{zz}$); (**k–n**) lateral displacements; (**p–s**) lateral strains ($s_{xx}$); (**e,o**) color bar in millimeters related to the figures in the same row; (**j,t**) color bar in percentages related to the figures in the same row. The axis of the displacement and gradient fields represents the position below the transducer in millimeters.

In Figure 7, the performances of the axial displacement estimates are summarized for all transducers, beam-forming strategies (including angular weighting and apodization) and line densities. As can be noticed for L7-4 (Figure 7a), the RMSE seemed to be almost constant for all beam-forming strategies and line-densities except DaS with one line per pitch, in which the RMSE increased. For the 12L4VF (Figure 7b), three lines per pitch provided the lowest RMSE using Lu's-fk, Stolt's-fk or DaS, with apodization or angular weighting. Furthermore, apodization or angular weighting seemed to decrease the RMSE for this transducer. However, the opposite effect of weighting and apodization can be seen for both the L12-5 and MS250 (Figure 7c,d) in which the RMSE increased. Multiple lines seemed to slightly increase the RMSE compared to one line per pitch for all strategies in these transducers. DaS or Lu's-fk with one line per pitch resulted in the lowest RMSE in both L12-5 and MS250. The RMSE by the lateral displacements estimates (Figure 8) decreased for each method and transducer using over 1 line per pitch.

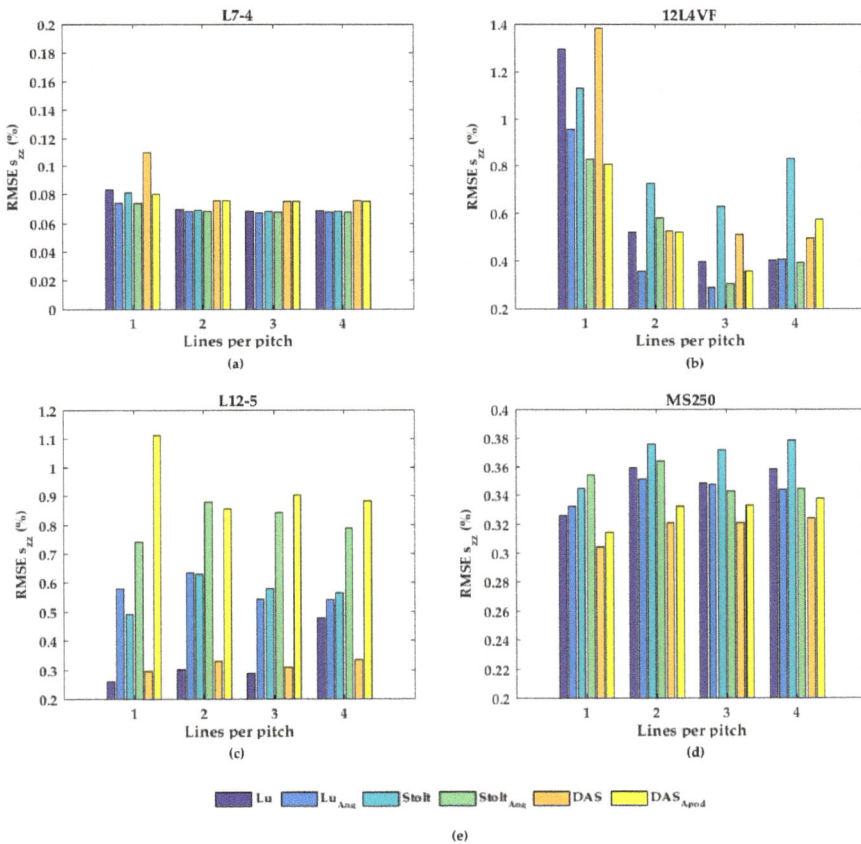

**Figure 7.** The root-mean-squared error (RMSE) of the axial strains ($s_{zz}$) for all beam-forming strategies and line densities for the (**a**) L7-4, (**b**) 12L4VF, (**c**) L12-5, and (**d**) MS250 transducer; (**e**) is the legend used in (**a**–**d**).

The RMSE increased when applying angular weighting or apodization for all transducers except 12L4VF, for which the RMSE decreased. Lu's-fk resulted in slightly decreased RMSE compared to other methods in both L7-4 and L12-5; Lu's-fk with angular weighting in 12L4VF, and Lu's-fk and DaS in MS250. For the 12L4VF, the RMSEs were recalculated neglecting the left top area (squared area; axial < 28 mm; lateral < 0 mm, Figure 6g,q) including the artifact. The results are presented in Figure 9. Although the overall RMSE values decreased, the observed results were similar to those for the full FoV (Figures 7b and 8b). Line densities of two or three lines per pitch resulted in the lowest RMSE for all strategies. Apodization or weighting seemed to decrease RMSE values but less severely and at times even increased values. This artifact might be caused by a small number of damaged elements on the left side of the transducer, resulting in artifacts especially in the near-field as only a few elements were used for reconstruction in that area.

Examples of lateral displacement and strain fields (L12-5, Lu's-fk) with a line density of one and two lines per pitch are shown in Figure 10. In the gradient field at 1 line per pitch (Figure 10b), bands with increased gradients are visible at axial positions of 20, 30, 40, and 50 cm. At these positions, the lateral displacements were approximately ±0.1 and ±0.3 mm, which were displacement at half-pitch positions. After beam-forming using two lines per pitch, these bands were decreased (Figure 10c,d). These bands resulted in an increased RMSE for one compared to multiple lines per pitch (Figure 10c). The appearance of these bands at sub-pitch displacements using one line per pitch were also observed for the other transducers and beam-forming methods.

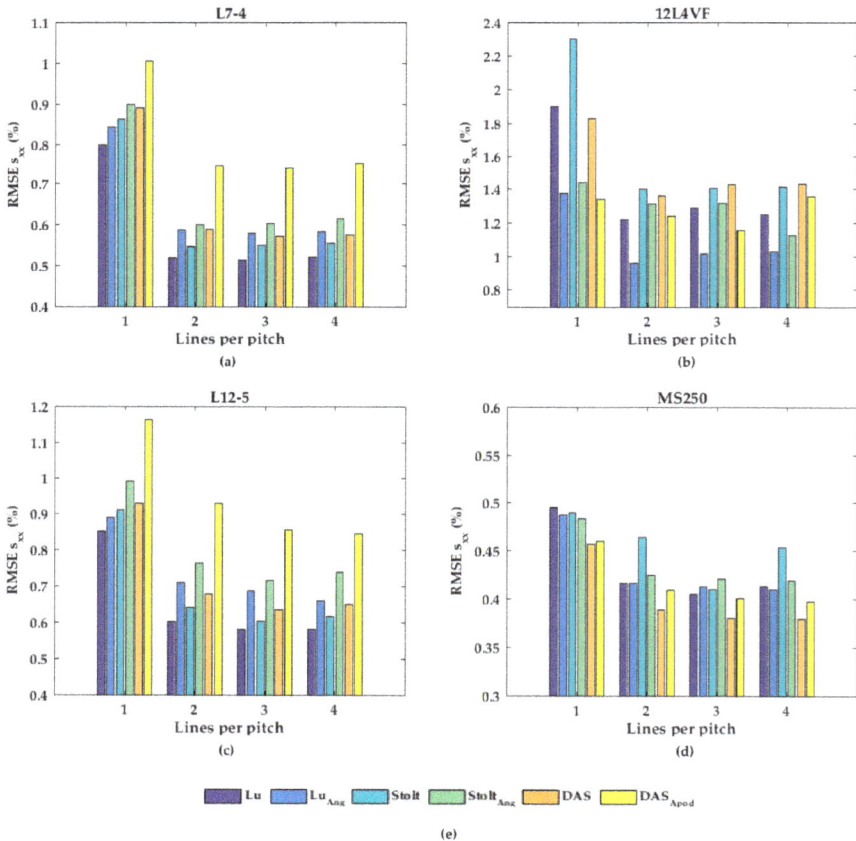

**Figure 8.** The root-mean-squared error (RMSE) of the lateral strains ($s_{xx}$) for all beam-forming strategies and line densities for the (**a**) L7-4, (**b**) 12L4VF, (**c**) L12-5, and (**d**) MS250 transducer; (**e**) is the legend used in (**a**–**d**).

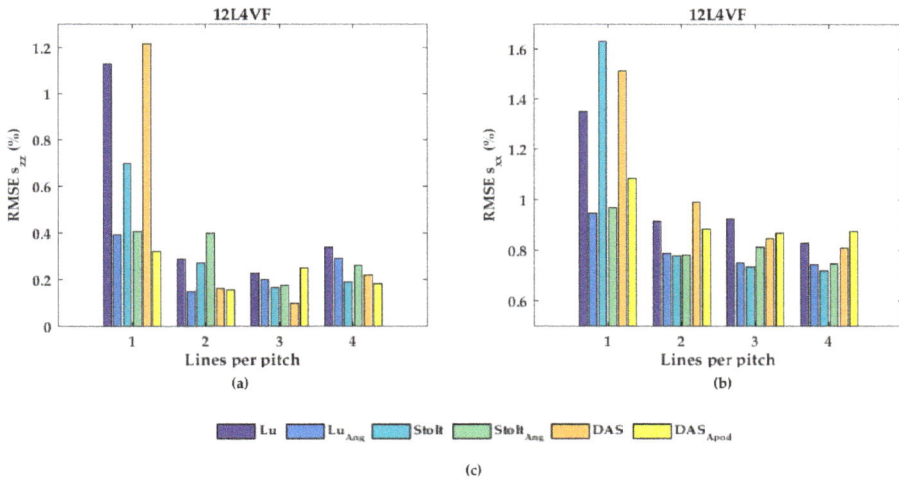

**Figure 9.** The root-mean-squared error (RMSE) without the top left artifact of the (**a**) axial strain ($s_{zz}$) and (**b**) lateral strain ($s_{xx}$) of the 12L4VF transducer; $s_{zz}$ and $s_{xx}$ values in the top left corner (axial and lateral position smaller than 28 and 0 cm respectively) were neglected in these RMSE calculations.

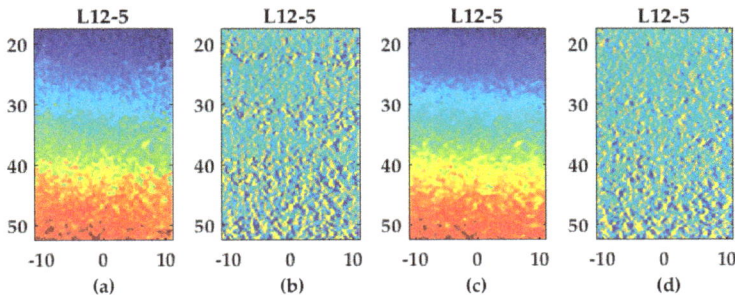

**Figure 10.** Lateral displacement and strain fields using Lu's-fk for the L12-5 transducer: (**a**,**c**) displacement field using one (**a**) or two (**c**) lines per pitch. Please refer to Figure 6o for the color bar; (**b**,**d**) strain field using one (**b**) or two (**d**) lines per pitch. Please refer to Figure 6j and Figure 6t for the corresponding color bars.

## 4. Discussion

In this study, the effect of beam-forming strategies and line density on the accuracy of normalized cross-correlation-based displacement estimation was evaluated. Therefore, we designed a method to evaluate the accuracy by phantom rotation, which had two main advantages. First, displacement fields were induced with axial and lateral displacements separately varying within ±0.5 mm (Figure 1a–d, i–l) instead of pure axial or lateral displacements. Second, the local displacement gradients are expected to be zero independent of the rotation center and angle in this case where the rotation angles was small (<<10°). Consequently, the error of the gradient (RMSE) was used as measure for the accuracy of displacement estimates.

Displacements were calculated by normalized cross-correlation and block-matching. Although other methods are available, only this method was used in the study since it is widely used in (shear) strain imaging. The effect of different methods on the accuracy of displacements was not the scope of this study and so not evaluated. Furthermore, a one dimensional, three-point least-square-strain estimator (LSQSE) was implemented to calculate the gradients for evaluation. Although not presented here, equal results were obtained for LSQSE sizes of 7 and 11 points.

The accuracy of displacement estimates, especially lateral, improved significantly using a line-density of multiple lines per pitch for all transducers except the high frequency transducer MS250 in which this improvement was minimal or the accuracy even slightly decreased for some beam-forming strategies. Improvement by higher line-densities was also observed by Konofagou et al. [20] in which conventional focused ultrasound was used and RF-lines were interpolated (up to 64 lines per pitch) to gain higher line-densities to improve results. The accuracy of displacement estimates already improved by two or three lines per pitch in this study. However, interpolation of the cross-correlation peak after the final iteration was implemented in this study which also contributes to improved sub-sample displacement estimates [2]. As can be seen in Figure 10a,b, the performance of cross-correlation peak interpolation seemed to be decreased in displacements at half-pitch distances resulting in increased gradient bands (Figure 10b). This effect seemed to disappear at a line density of two lines per pitch (Figure 10c,d). The increased performance of sub-sample interpolation might be explained by the point spread function (PSF). According to Saris et al. [17], beam-forming to a PSF-based USGrid results in a circular cross-correlation peak which improves the performance of two-dimensional spline peak interpolation. The optimal number lines should be between two or three lines according to the point spread function for all transducers. This line-density matched the optimal density found in this study.

For all transducers except the 12L4VF transducer, apodization or angular weighting seemed to decrease or minimally affect the accuracy of axial and lateral displacement estimates (Figures 7 and 8). Weighting and apodization increased the lateral resolution of the beam-formed data (Appendix A), which may explain the decreased accuracy of lateral estimates. In Appendix A, the axial and lateral resolution and contrast-to-noise-ratios (CNR) of the beam-formed data by the different beam-forming, weighting and apodization methods were evaluated. The relation between lateral resolution and displacement estimation accuracy was also described previously by Luo et al. [16]. Since lateral and axial displacement estimation was coupled, the accuracy of lateral displacement estimates also affected the accuracy of the axial estimates. For the 12L4VF transducer, the opposite effect can be noticed in the results with and without the top left artifact (Figures 7b, 8b and 9). The increase in performance with apodization or angular weighting may be partially explained because weighting or apodization reduced artifacts in beam-formed data caused by broken elements in the left part of the transducer footprint. As can be observed in Figure 9, the improvement in displacement estimation accuracy by weighting was less when neglecting the artifact. Another explanation might be related to the fact that side lobes are stronger for the 12L4VF compared to the other transducers given its increased element width-to-wavelength ratio. Weighting suppresses side lobes which might also explain the improved displacement estimates despite the worse lateral resolution. In this study, a homogenous phantom was used so the effect of improved contrast (Appendix A) by weighting might be limited and has to be further investigated in an inhomogeneous phantom or in ex vivo tissue which can also be rotated. Furthermore, Lu's-fk performed equally or outperformed DaS with optimal line density and for all transducers. This implied that the computationally more efficient Lu's-fk method can be used as alternative for the widely applied DaS. Furthermore, Lu's-fk often outperformed Stolt's-fk method probably because Lu's-fk resulted in an improved lateral resolution in the beam-formed data (Appendix A).

## 5. Conclusions

In this study, we investigated the effect of different beam-forming strategies on the accuracy of displacement estimates in four different transducers (L7-4, 12L4VF, L12-5 and MS250). We developed an easy-to-implement method to evaluate the displacement estimation accuracy based on the displacement gradients estimated in a rotating phantom. A line density of multiple lines per

*Appl. Sci.* **2018**, *8*, 319

pitch seemed to outperform the accuracy of the estimates compared to one line per pitch for all beam-forming strategies and transducers. Lu's-fk beam-forming seemed to outperform Stolt's-fk and resulted in similar accuracy as DaS for multiple lines per pitch. It can therefore be used interchangeable in displacement estimation without affecting accuracy, with the added advantage that the computational load is lower. In future work, an inhomogeneous phantom or ex vivo tissue might be used to improve extrapolation of results to in vivo conditions and to further investigate the effect of contrast enhancement and side lobe reduction by apodization or weighting strategies on displacement estimation.

**Acknowledgments:** This research is supported by Siemens Healthineers and by the Dutch Technology Foundation STW (Project 13290) which is part of the Netherlands Organization for Scientific Research (NWO) and is partly funded by the Ministry of Economic Affairs. Furthermore, the authors wish to thank Anne Saris in our group for all constructive discussions and Chi-Yin Lee, John Klepper and Andy Milkowski of Siemens Healthineers for their technical support.

**Author Contributions:** G.A.G.M.H. and H.H.G.H. conceived and designed the experiments; G.A.G.M.H. performed the experiments; G.A.G.M.H., C.C. and H.H.G.H. implemented the beam-forming and displacement estimation algorithms; G.A.G.M.H., C.C., H.H.G.H. and C.L.d.K. analyzed the data; G.A.G.M.H., C.C., H.H.G.H. and C.L.d.K. wrote the paper.

**Conflicts of Interest:** The authors declare no conflict of interest.

## Appendix A

**Introduction:** Element data obtained by plane-wave acquisitions were beam-formed to reconstruct radio frequency (RF) data for displacement estimation. In this study, three beam-forming strategies were used: delay-and-sum (DaS), Lu's-fk, and Stolt's-fk. In DaS beam-forming, dynamic focusing in receive was applied with and without apodization (Hamm function) in which the F-number had to be set (Equation (1)). Furthermore, angular weighting was applied in Lu's-fk and Stolt's-fk beam-forming in which the migrated k-space spectrum was multiplied with a template designed such that wave directions closer aligned to the beam-steering direction were multiplied with higher weights compared to waves deviating from the steering angle. In angular weighting, the angular range had to be set: wave directions within this range were weighted by Hann function and outside were weighted by zero. Summarized, the optimal F-number in DaS with and without apodization and the range in angular weighting were investigated.

**Method and Materials:** Plane-wave element data were obtained by three transducers (L7-4, 12L4VF, L12-5) in a multi-purpose phantom (model 539; ATS Laboratories, Norfolk, VA, USA) containing circular lesions and needles to evaluate contrast and resolution, respectively. Element data were beam-formed by above strategies in which the F-number (DaS) or angular range (angular weighting) were varied between 0 and 2.0, and ±10 and ±30. The mean contrast-to-noise ratio (CNR) of the lesions and mean axial and lateral resolution (full width at half maximum) of the needles were calculated to a depth of 50 mm (Figure A1) using software provided by the Plane-wave Imaging Challenge in Medical Ultrasound (PICMUS) of the IEEE International Ultrasonics Symposium 2016 [21].

**Results and Discussion:** In DaS, an F-number of 0.875 seemed to be most optimal for all transducers: contrast decreased while lateral resolution remained constant at values below 0.875 and contrast remained constant above 0.875 while lateral resolution increased. The axial resolution seemed to be independent of the F-number. These results were found in DaS with and without apodization. As can be seen in Table A1, apodization increased contrast, which came at the cost of increased lateral resolution.

Similar results were found for angular weighting (Table A2), a range of ±20° seemed to be optimal for all transducers and both Lu's-fk and Stolt's-fk, since the lateral resolution and contrast did not improve anymore at larger and smaller angle ranges, respectively. The axial resolution was unaffected by the angular range. Furthermore, angular weighting improved contrast at the cost of lateral resolution, as can be seen in comparing Tables A2 and A3.

The multi-purpose phantom used in this study was only suitable for transducers with frequencies varying between 2 and 15 MHz. Therefore, the MS250 transducer with a center frequency of 21 MHz could not be evaluated and the same F-number and angular range were used as for the other transducers.

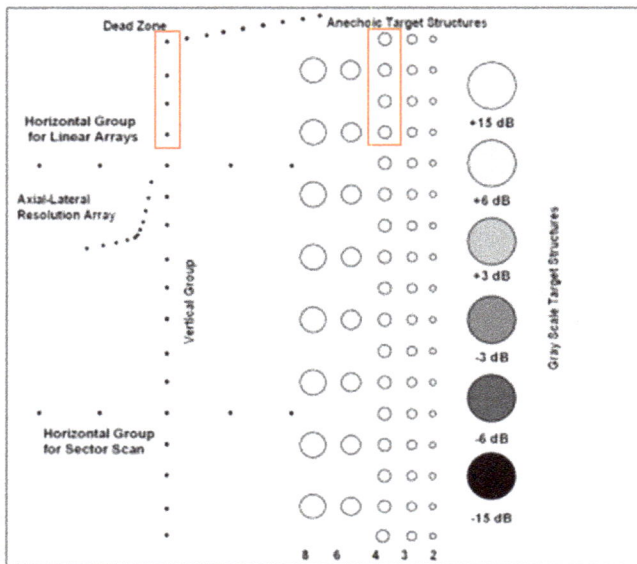

**Figure A1.** Cross-section of the multi-purpose phantom and the needles and lesions used for analysis indicated by the orange rectangular boxes.

**Table A1.** Axial and lateral resolution, and contrast-to-noise ratio (CNR) by DaS (F-number of 0.875).

| Transducer | Apodization | Ax. Res. | Lat. Res | CNR |
|------------|-------------|----------|----------|--------|
| L7-4 | No apod | 421 μm | 490 μm | 7.8 dB |
| - | Hamm | 422 μm | 651 μm | 9.7 dB |
| 12L4VF | No apod | 422 μm | 426 μm | 5.4 dB |
| - | Hamm | 419 μm | 577 μm | 7.4 dB |
| L12-5 | No apod | 406 μm | 426 μm | 5.3 dB |
| - | Hamm | 405 μm | 550 μm | 7.2 dB |

**Table A2.** Axial and lateral resolution, and CNR by angular weighted Lu's-fk and Stolt's-fk (range ±20°).

| Transducer | Method | Ax. Res. | Lat. Res | CNR |
|------------|-----------|----------|----------|--------|
| L7-4 | Lu's-fk | 423 μm | 601 μm | 9.5 dB |
| - | Stolt's-fk | 423 μm | 610 μm | 9.5 dB |
| 12L4VF | Lu's-fk | 406 μm | 530 μm | 6.9 dB |
| - | Stolt's-fk | 408 μm | 616 μm | 6.7 dB |
| L12-5 | Lu's-fk | 400 μm | 510 μm | 7.4 dB |
| - | Stolt's-fk | 401 μm | 556 μm | 7.4 dB |

**Table A3.** Axial and lateral resolution, and CNR by Lu's-fk and Stolt's-fk without angular weighting.

| Transducer | Method | Ax. Res. | Lat. Res | CNR |
|---|---|---|---|---|
| L7-4 | Lu's-fk | 420 μm | 493 μm | 8.1 dB |
| | Stolt's-fk | 416 μm | 507 μm | 7.7 dB |
| 12L4VF | Lu's-fk | 405 μm | 423 μm | 5.4 dB |
| | Stolt's-fk | 412 μm | 584 μm | 4.7 dB |
| L12-5 | Lu's-fk | 397 μm | 420 μm | 6.3 dB |
| | Stolt's-fk | 402 μm | 532 μm | 5.6 dB |

## References

1. Ophir, J.; Céspedes, I.; Ponnekanti, H.; Yazdi, Y.; Li, X. Elastography: A quantitative method for imaging the elasticity of biological tissues. *Ultrason. Imaging* **1991**, *13*, 111–134. [CrossRef] [PubMed]
2. Lopata, R.G.; Nillesen, M.M.; Hansen, H.H.; Gerrits, I.H.; Thijssen, J.M.; de Korte, C.L. Performance evaluation of methods for two-dimensional displacement and strain estimation using ultrasound radio frequency data. *Ultrasound Med. Boil.* **2009**, *35*, 796–812. [CrossRef] [PubMed]
3. Chen, H.; Varghese, T. Multilevel hybrid 2d strain imaging algorithm for ultrasound sector/phased arrays. *Med. Phys.* **2009**, *36*, 2098–2106. [CrossRef] [PubMed]
4. D'Hooge, J.; Heimdal, A.; Jamal, F.; Kukulski, T.; Bijnens, B.; Rademakers, F.; Hatle, L.; Suetens, P.; Sutherland, G.R. Regional strain and strain rate measurements by cardiac ultrasound: Principles, implementation and limitations. *Eur. J. Echocardiogr.* **2000**, *1*, 154–170. [CrossRef] [PubMed]
5. Luo, J.; Konofagou, E. A fast normalized cross-correlation calculation method for motion estimation. *IEEE Trans. Ultrason. Ferroelectr. Freq. Control* **2010**, *57*, 1347–1357. [PubMed]
6. Pan, X.; Gao, J.; Tao, S.; Liu, K.; Bai, J.; Luo, J. A two-step optical flow method for strain estimation in elastography: Simulation and phantom study. *Ultrasonics* **2014**, *54*, 990–996. [CrossRef] [PubMed]
7. Kallel, F.; Ophir, J. A least-squares strain estimator for elastography. *Ultrason. Imaging* **1997**, *19*, 195–208. [CrossRef] [PubMed]
8. Delannoy, B.; Torguet, R.; Bruneel, C.; Bridoux, E. Ultrafast electronical image reconstruction device. In *Echocardiology*; Lancée, C.T., Ed.; Springer Netherlands: Dordrecht, The Netherlands, 1979; pp. 447–450.
9. Delannoy, B.; Torguet, R.; Bruneel, C.; Bridoux, E.; Rouvaen, J.M.; Lasota, H. Acoustical image reconstruction in parallel-processing analog electronic systems. *J. Appl. Phys.* **1979**, *50*, 3153–3159. [CrossRef]
10. Montaldo, G.; Tanter, M.; Bercoff, J.; Benech, N.; Fink, M. Coherent plane-wave compounding for very high frame rate ultrasonography and transient elastography. *IEEE Trans. Ultrason. Ferroelectr. Freq. Control* **2009**, *56*, 489–506. [CrossRef] [PubMed]
11. Lu, J.Y. 2D and 3D high frame rate imaging with limited diffraction beams. *IEEE Trans. Ultrason. Ferroelectr. Freq. Control* **1997**, *44*, 839–856. [CrossRef]
12. Lu, J.Y. Experimental study of high frame rate imaging with limited diffraction beams. *IEEE Trans. Ultrason. Ferroelectr. Freq. Control* **1998**, *45*, 84–97. [CrossRef] [PubMed]
13. Garcia, D.; Le Tarnec, L.; Muth, S.; Montagnon, E.; Poree, J.; Cloutier, G. Stolt's f-k migration for plane wave ultrasound imaging. *IEEE Trans. Ultrason. Ferroelectr. Freq. Control* **2013**, *60*, 1853–1867. [CrossRef] [PubMed]
14. Chen, C.; Hendriks, G.A.G.M.; Hansen, H.H.G.; de Korte, C.L. Design of an angular weighting template for coherent plane wave compounding in fourier domain. In Proceedings of the 2017 IEEE International Ultrasonics Symposium (IUS), Washington, DC, USA, 7–9 September 2017; p. 1.
15. Ophir, J.; Alam, S.K.; Garra, B.; Kallel, F.; Konofagou, E.; Krouskop, T.; Varghese, T. Elastography: Ultrasonic estimation and imaging of the elastic properties of tissues. *Proc. Inst. Mech. Eng. Part H J. Eng. Med.* **1999**, *213*, 203–233. [CrossRef] [PubMed]
16. Luo, J.; Konofagou, E.E. Effects of various parameters on lateral displacement estimation in ultrasound elastography. *Ultrasound Med. Boil.* **2009**, *35*, 1352–1366. [CrossRef] [PubMed]
17. Saris, A.E.C.M.; Nillesen, M.M.; Fekkes, S.; Hansen, H.H.G.; de Korte, C.L. Robust blood velocity estimation using point-spread-function-based beamforming and multi-step speckle tracking. In Proceedings of the 2015 IEEE International Ultrasonics Symposium (IUS), Taipei, Taiwan, 21–24 October 2015; pp. 1–4.

18.   Hansen, H.H.G.; Lopata, R.G.P.; Idzenga, T.; de Korte, C.L. Full 2D displacement vector and strain tensor estimation for superficial tissue using beam-steered ultrasound imaging. *Phys. Med. Boil.* **2010**, *55*, 3201–3218. [CrossRef] [PubMed]

19.   Fekkes, S.; Swillens, A.E.; Hansen, H.H.; Saris, A.E.; Nillesen, M.M.; Iannaccone, F.; Segers, P.; de Korte, C.L. 2D versus 3D cross-correlation-based radial and circumferential strain estimation using multiplane 2d ultrafast ultrasound in a 3d atherosclerotic carotid artery model. *IEEE Trans. Ultrason. Ferroelectr. Freq. Control* **2016**, *63*, 1543–1553. [CrossRef] [PubMed]

20.   Konofagou, E.; Ophir, J. A new elastographic method for estimation and imaging of lateral displacements, lateral strains, corrected axial strains and poisson's ratios in tissues. *Ultrasound Med. Biol.* **1998**, *24*, 1183–1199. [CrossRef]

21.   Liebgott, H.; Rodriguez-Molares, A.; Cervenansky, F.; Jensen, J.A.; Bernard, O. Plane-wave imaging challenge in medical ultrasound. In Proceedings of the 2016 IEEE International Ultrasonics Symposium (IUS), Tours, France, 18–21 September 2016; pp. 1–4.

*applied*
*sciences*

MDPI

*Article*

# High-Frame-Rate Doppler Ultrasound Using a Repeated Transmit Sequence

**Anthony S. Podkowa** [1,*], **Michael L. Oelze** [1] **and Jeffrey A. Ketterling** [2]

[1] Beckman Institute, University of Illinois at Urbana-Champaign, Urbana, IL 61801, USA; oelze@illinois.edu
[2] Lizzi Center for Biomedical Engineering, Riverside Research Institute, New York, NY 10038, USA; Jketterling@RiversideResearch.org
* Correspondence: tpodkow2@illinois.edu; Tel.: +1-815-556-2729 or +1-815-954-4873

Received: 5 December 2017; Accepted: 28 January 2018; Published: 1 February 2018

**Abstract:** The maximum detectable velocity of high-frame-rate color flow Doppler ultrasound is limited by the imaging frame rate when using coherent compounding techniques. Traditionally, high quality ultrasonic images are produced at a high frame rate via coherent compounding of steered plane wave reconstructions. However, this compounding operation results in an effective downsampling of the slow-time signal, thereby artificially reducing the frame rate. To alleviate this effect, a new transmit sequence is introduced where each transmit angle is repeated in succession. This transmit sequence allows for direct comparison between low resolution, pre-compounded frames at a short time interval in ways that are resistent to sidelobe motion. Use of this transmit sequence increases the maximum detectable velocity by a scale factor of the transmit sequence length. The performance of this new transmit sequence was evaluated using a rotating cylindrical phantom and compared with traditional methods using a 15-MHz linear array transducer. Axial velocity estimates were recorded for a range of ±300 mm/s and compared to the known ground truth. Using these new techniques, the root mean square error was reduced from over 400 mm/s to below 50 mm/s in the high-velocity regime compared to traditional techniques. The standard deviation of the velocity estimate in the same velocity range was reduced from 250 mm/s to 30 mm/s. This result demonstrates the viability of the repeated transmit sequence methods in detecting and quantifying high-velocity flow.

**Keywords:** color flow doppler; high-frequency ultrasound; plane-wave imaging; Nyquist velocity; multirate signal processing

## 1. Introduction

Doppler ultrasound is widely used to estimate velocity of blood flow and tissue motion in biological specimens [1]. Traditional Doppler approaches estimate the axial component of motion, where the maximum velocity is limited by the pulse-repetition frequency (PRF) of the ultrasound transmission as described by the Doppler equation (see Equation (9) in Section 2.1.5). In the most common estimation algorithms, such as the lag-one autocorrelation approach [2], velocities that exceed this limit result in phase wrapping (i.e., aliasing) and can result in estimates in the opposite direction of the actual flow. However, using conventional ultrasonic methods, if the Doppler estimation occurs throughout an image frame, such as in color flow Doppler, then the maximum velocity is now limited by the frame rate.

With conventional ultrasonic methods, images are formed by using one focused transmit per lateral scan line. This design decision, while simple to implement and well understood, results in a temporal bottleneck on the frame rate. Because the total acquisition time for each transmit event is the round-trip time of flight to the maximum depth of interest and back, the total image acquisition time is scaled by the number of transmit events (on the order of 128 or more). Thus, if high frame rate images are desired, reducing the number of transmit events for each frame is paramount.

Coherent compounding approaches were introduced to minimize the number of transmit events for each image frame and thereby increase the absolute frame rate without sacrificing image quality. These approaches include limited diffraction beams [3–5], steered plane waves [6], and diverging waves [7,8]. Each of these methods produces high-quality images at frame rates in the kHz range. Acquiring data at the highest frame rate is paramount for accurate reconstruction of high velocity motion, such as that observed in mouse blood flow, where velocities can be as high as 900 mm/s [9]. This is particularly true for high-frequency ultrasound (>20 MHz) because, based on the Doppler equation, an increase in transmit frequency will lower the maximum velocity that can be resolved before aliasing.

Compounding approaches result in an effective slow-time downsampling of the data, thus reducing the frame rate by a factor of $N_\phi$, where $N_\phi$ is the number of angles transmitted per compounded frame. In addition, because the plane-wave compounding approach relies on a stationary medium assumption, the compounding process can result in signal loss when particles move a half-integer multiple of a wavelength over the compounding interval [10]. For this reason, using the precompounded data is advantageous for high velocity imaging. This idea has been successfully exploited to compensate for motion artifacts in coherent compounding techniques [11,12].

Use of the precompounded data for Doppler estimation comes with its own costs. Because each precompounded image is transmitted at a different steering angle, sidelobes in the point spread function (PSF) rotate about the target location even when the target is stationary. This sidelobe motion results in a periodic noise source which can corrupt velocity estimates. In order to address this issue, we propose using a repeated transmission sequence to avoid sidelobe motion on a short time scale. Specifically, we propose two new approaches which use this new transmit sequence. The first method estimates flow by evaluating the slow-time autocorrelation between adjacent precompounded images at the same transmit steering angle and aggregating the individual autocorrelation estimates over all angles. The second approach generates compounded frames for the initial and repeat events, respectively, and correlates these two frames. For both approaches, the autocorrelation occurs at a lag equivalent to one transmit event as opposed to a full compounded frame, resulting in an increase of the Nyquist velocity on the order of the number of angles used in the transmit sequence.

The proposed techniques were validated by using a cylindrical hyperechoic phantom mounted on a motor rotated at a known angular velocity. Data were acquired using two traditional linear transmit sequences of different lengths as well as the new double transmit sequence. For each acquired dataset, the received in-phase and quadrature (IQ) data were beamformed, and velocities were estimated using both the proposed and traditional methods for both the compounded and precompounded datasets. Power Doppler and color flow Doppler images were generated, and the root-mean-square error, standard deviation, and bias were evaluated for each processing scheme.

## 2. Materials and Methods

### 2.1. Theory

#### 2.1.1. Coherent Plane-Wave Compounding

A plane-wave transmission is generated by firing all of the elements of a linear array with inter-element transmission delays that create a coherent wave front at a steered angle. The return echoes are then digitized on all of the array elements and beamformed data representing a full image frame can be obtained with delay-and-sum methods. Plane waves are usually transmitted in a sequence of $N_T$ steered excitations spanning a set of $N_\phi$ distinct angles, where the angle of transmission linearly increases from low to high (Figure 1a). Note that in the traditional case $N_T = N_\phi$. The data collected from each transmit event, after beamforming, are coherently summed to form high-resolution images (HRIs), which improves image resolution and contrast [6].

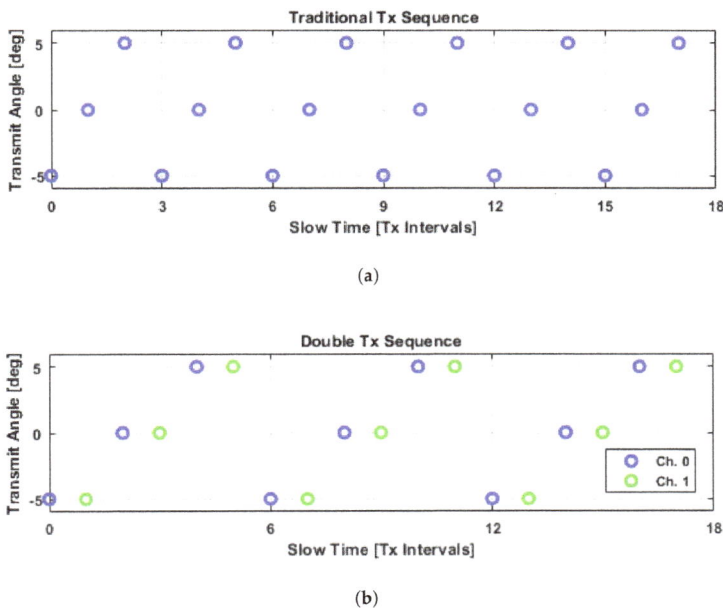

**Figure 1.** Comparison of the different transmit sequences used in this study. With the traditional transmit sequence, acquisitions with the same point spread function are spaced apart over intervals equivalent to a transmit sequence length. Due to the fact that these components will exhibit the strongest spatial correlations, reordering them such that they are temporally adjacent allows for correlation estimates at the shortest possible time interval. (**a**) Traditional transmit sequence where the transmit angles linearly increment and then repeat; (**b**) Double transmit sequence where the transmit angle repeats before linearly incrementing to the next angle.

The fundamental tradeoff when using a coherent compounding technique is between the spatial and temporal resolution. As noted in [13], the effect of the coherent compounding process is one of a temporal low-pass, finite impulse response (FIR) filter. However, Ekroll, et al. [13], did not address the effect of the downsampling operation. As described by multirate signal processing theory [14,15], a downsampling operation results in an effective expansion of the discrete time spectrum, which can result in additional aliasing. Because traditional coherent compounding techniques use uniform weights [6], the anti-aliasing filter corresponds to a rectangular filter (See Appendix) that is prone to temporal frequency sidelobes. Equally weighting the compounded components could lead to slow-time temporal aliasing and high velocity estimates would be particularly vulnerable to this source of error.

In principle, one could design an anti-aliasing filter with better stop band characteristics using traditional signal processing techniques [15]. However, such a design process would inevitably produce a nonuniformly weighted low pass filter, and as such would result in reduced signal contribution from the highly steered components of the compounding process. Consequently, the spatial resolution of the compounded images would be reduced.

Another approach would be to use precompounded low-resolution images (LRIs). However, because each of the steered plane waves produces a slightly different PSF, a new temporal noise source is introduced in the form of spatial sidelobe motion that oscillates periodically in slow time (Figure 2). This sidelobe motion could be mistaken as tissue motion by traditional Doppler estimation techniques [2], especially in hypoechoic tissue regions. While in principle such sidelobes could be minimized via an appropriately designed apodization function, such a design choice ultimately results in a loss of spatial resolution. As such, it is important to consider alternative processing schemes to address this new noise source associated with LRI-based Doppler estimation.

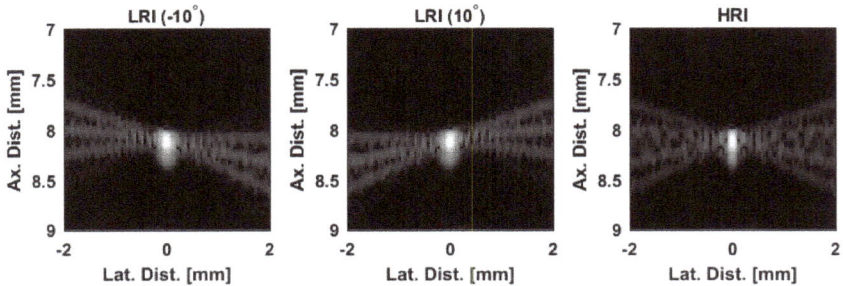

**Figure 2.** The coherent compounding approach generates high-resolution images (HRIs)by summing together low-resolution images (LRIs) formed at different steering angles. If we look at each LRI individually, we can see that the point spread function (PSF) rotates about the center as the steering angle changes. If we use the LRIs to produce Doppler estimates, the sidelobe rotation could be misinterpreted as tissue motion by the lag-one autocorrelation algorithm. Images are rendered at a 60-dB scale and are normalized to their corresponding maxima.

### 2.1.2. The Polyphase Decomposition

A potential alternative to resolve this issue is to decompose the slow-time signal into its $N_T$ polyphase components [14]. In the field of multirate signal processing, an $N$th order polyphase decomposition of a discrete signal $s[n]$ is a set of $N$ periodically interleaved subsequences sampled at $1/N$ times the framerate. Mathematically, the $k$th polyphase component signal $e_k[n]$ is given by

$$e_k[n] = s[Nn + k] \tag{1}$$
$$k \in \{0 : N - 1\}. \tag{2}$$

If we preserve all the polyphase components, the original signal can be reconstructed exactly by interleaving, such that

$$s[n] = e_{\mathrm{mod}(n,N)}\left[\left\lfloor \frac{n}{N} \right\rfloor\right], \tag{3}$$

where $\mathrm{mod}(n, N)$ is the remainder after division of $n$ by $N$ and $\lfloor \cdot \rfloor$ is the floor function.

Noting that if we choose the polyphase decomposition order $N$ such that $N = N_T$, each polyphase component in the slow-time signal corresponds to a particular angle in the transmit sequence. Direct comparisons can be made between slow-time signals of the same transmit steering angle, and thus would be insensitive to sidelobe motion. However, if we choose to do so with a conventional linear sequence, there is no performance improvement relative to just using the HRIs, as each polyphase channel operates at the HRI frame rate. For this reason, we introduce the double transmit sequence as a workaround, as examined in the next subsection.

### 2.1.3. Double Transmit Sequence

In order to mitigate the effect of sidelobe motion in the precompounded LRIs, a repeated transmit sequence is proposed (Figure 1b). By repeating the transmit angle each time, motion can be analyzed between frames with the same PSF at a shorter time interval than would be possible with a traditional linearly incrementing transmit sequence. Such an approach opens up the design space to new Doppler estimation approaches (Sections 2.1.6 and 2.1.7).

One consequence of this design decision is that it doubles the length of the transmit sequence such that $N_T = 2N_\phi$ instead of $N_\phi$ as before. This means that if we choose to compound the sequence to generate HRIs for review, the HRI frame rate will be half that of the linear transmit sequence. However, given that the maximum velocity range is increased by a factor of $N_\phi$ as described in Sections 2.1.6 and 2.1.7, and that typically one averages over multiple HRI frames anyway, one can argue that the decision is justifiable. That being said, the efficacy of such an approach should be evaluated by the end user to determine if tradeoff is acceptable.

## 2.1.4. Clutter Filtering

Prior to estimating velocity, a clutter filter is typically used to suppress stationary components of the slow-time signal. Traditionally this is accomplished with a high-pass filter. In the case of HRIs, the PSF of each frame is the same and, therefore, clutter filtering may be handled with standard digital filtering techniques [15]. However, because the PSF in each LRI frame varies with time, the options for clutter filtering become more varied. If we wish to only compare images with the same PSF during the clutter filtering process, we must separate our slow-time signal into polyphase channels with the same steering angle and process them independently. This separation results in $N_T$ different polyphase channels operating at a slow-time sampling rate equivalent to that of the HRIs. Using the principles of multirate signal processing, one can show that compounding the LRIs after filtering this way will result in the same digital frequency response as a simple clutter filter on the HRIs (see Theorem A2 in the Appendix). This result is a direct consequence of one of the noble identities of multirate signal processing [14].

It is important to note that using such a filtering scheme comes with some significant drawbacks. Because the clutter filtering occurs at the HRI frame rate, the frequency response of the filter repeats $N_T$ times when analyzed with respect to the LRI frame rate. This results in stop bands occurring at integer multiples of the HRI frame rate. Therefore, velocity estimates within these slow-time bands will be less consistent than at other frequencies.

## 2.1.5. Lag-One Estimation

In order to estimate the axial motion at each pixel in the image, a lag-one autocorrelation approach was used [2]. Using the beamformed, complex analytic signal $s_a(\vec{r}, t)$ at each pixel location, the complex slow-time autocorrelation was evaluated at a single sample lag via conjugate product followed by convolution with a rectangular filter, yielding

$$R_1(\vec{r}, t, \tau = \Delta t) = \mathrm{rect}\left(\frac{t}{T}\right) \underset{t}{*} \left(s_a^*(\vec{r}, t) s_a(\vec{r}, t + \Delta t)\right), \tag{4}$$

where $R_1(\vec{r}, t, \tau)$ is the autocorrelation estimate at location $\vec{r}$, $t$ is the slow-time instant, $\tau$ is the lag, $T$ is the averaging interval, and $\Delta t$ is the sampling interval. The instantaneous phase shift is then estimated via an arctangent and converted to velocity via the Doppler equation, yielding

$$\hat{\phi}(\vec{r}, t) = \mathrm{atan2}\left(\mathrm{Im}\{R_1(\vec{r}, t, \Delta t)\}, \mathrm{Re}\{R_1(\vec{r}, t, \Delta t)\}\right) \tag{5}$$

$$\hat{\omega}(\vec{r}, t) = \frac{\hat{\phi}(\vec{r}, t)}{\Delta t} \tag{6}$$

$$\hat{v}_z(\vec{r}, t) = \frac{\hat{\omega}(\vec{r}, t)}{2\pi f_0} \frac{c}{2} \tag{7}$$

where $\hat{\phi}$ is the phase estimate in radians, $\hat{\omega}$ is the instantaneous slow-time frequency estimate, $\hat{v}_z$ is the velocity estimate, $f_0$ is the transmitted center frequency, and $c$ is the sound speed of the medium.

Due to the phase ambiguity of the arctangent function, possible values for $\hat{\phi}$ are limited to the interval $(-\pi, \pi]$. Because of this ambiguity, the phase-based velocity estimation algorithms exhibit aliasing when the velocity magnitude exceeds the slow-time Nyquist rate, given by

$$v_N \equiv \max \hat{v}_z = \frac{f_N}{f_0} \frac{c}{2} \tag{8}$$

$$= \frac{1}{2\Delta t f_0} \frac{c}{2} \tag{9}$$

where $f_N = 1/(2\Delta t)$ is the Nyquist frequency. For this reason, it is advantageous to utilize the smallest sampling interval possible when estimating high velocities using high-frequency ultrasound.

### 2.1.6. Pairwise Lag-One Estimation

The lag-one estimation algorithm described above is versatile enough to be used with either compounded or precompounded images, with the latter being advantageous from a Nyquist standpoint, as it allows for a shorter sampling interval. However, using the lag-one estimation algorithm on the LRIs may result in artifacts due to the rotation of the PSF. By repeating each angle immediately after transmission, we can eliminate these artifacts while simultaneously maintaining the short slow-time sampling interval characteristic of the LRIs. This leads to a variant of the lag-one estimation algorithm which we shall call pairwise lag-one estimation. In pairwise lag-one estimation, we form the short-time autocorrelation estimates only on adjacent LRI pairs which share the same transmission angle (and therefore the same PSF). For notational purposes, let us define the slow-time complex analytic signal corresponding to angle $\phi_i$, repeat channel $m$, and slow-time sample $n$ as

$$s_a(\vec{r}, m, \phi_i, n) = s_a(\vec{r}, t(m, i, n)) \tag{10}$$

where

$$t(m, i, n) = (nN_T + 2i + m)\Delta t \tag{11}$$
$$i \in \{0 : N_\phi - 1\} \tag{12}$$
$$m \in \{0 : 1\} \tag{13}$$
$$n \in \mathbb{Z} \tag{14}$$
$$\phi_i = i\Delta\phi + \phi_0 \tag{15}$$
$$N_T = 2N_\phi \tag{16}$$

and $N_\phi$ is the number of angles used in the transmission sequence.

Using this convention, we may now define a slow-time lag-one autocorrelation estimate for each angle via conjugate product such that

$$\hat{R}_2(\vec{r}, t = (2i + nN_T)\Delta t, \tau = \Delta t) = s_a^*(\vec{r}, 0, \phi_i, n)s_a(\vec{r}, 1, \phi_i, n). \tag{17}$$

By defining the autocorrelation this way, adjacent autocorrelation estimates are now temporally indexed at half the LRI sampling rate, but are still lagged by one LRI slow-time sample. This effectively allows for phase estimation at the LRI frame rate, but without any motion artifacts due to the rotation of the PSF. At this point, averaging may be performed at this half sample rate via a convolution with a rectangular filter yielding

$$R_2(\vec{r}, t, \Delta t) = \mathrm{rect}\left(\frac{t}{T}\right) \underset{t}{*} \hat{R}_2(\vec{r}, t, \Delta t). \tag{18}$$

From this point, the velocity estimation process is the same as given in Equations (5)–(7).

### 2.1.7. HRI Pairwise Lag-One Estimation

Yet another approach that can be pursued with this estimation scheme is using an interleaved compounding scheme. In this case, we generate two HRI channels corresponding to the value of $m$, given by

$$h_a(\vec{r}, m, n) = \sum_{i=0}^{N_\phi - 1} s_a(\vec{r}, m, \phi_i, n). \tag{19}$$

In the case depicted in Figure 1b, the coherent compounding produces two HRIs per transmit sequence period, one each for the blue and green subsets. Each compounded frame is thus made up of samples spaced at an interval of $2\Delta t$, but each component of these HRI channels has a lag of $\Delta t$ relative to its partner. This way, we can boost the signal for slowly moving points in the field, but still have relative lags at the LRI time scale. The lag-one autocorrelation estimate is then given by

$$\hat{R}_3(\vec{r}, t = nN_T\Delta t, \tau = \Delta t) = h_a^*(\vec{r}, 0, n)h_a(\vec{r}, 1, n) \tag{20}$$

$$R_3(\vec{r}, t = nN_T\Delta t, \tau = \Delta t) = \text{rect}\left(\frac{t}{T}\right) \underset{t}{*} \hat{R}_3(\vec{r}, t, \Delta t). \tag{21}$$

In this case the autocorrelation is sampled in intervals of $N_T\Delta t$ but again the lag is estimated at an interval of $\Delta t$, allowing for a higher Nyquist velocity. From this point, the velocity estimation process is the same as given in Equations (5)–(7).

### 2.2. Experimental Design

#### 2.2.1. Ultrasound Acquisition

Data were acquired with a Vantage 128 ultrasound research platform (Verasonics, Redmond, WA) and an 15.6-MHz, 128-element linear array (Verasonics L22-14v) (Table 1). Plane waves were transmitted at a PRF of 30 kHz using a Tukey cosine apodization with a 10% rolloff parameter. The transmit frequency was 17.857 MHz, the duration of the transmit waveform was 1.5 cycles, and the transmission angles ranged from −5 to 5 degrees. Received data were bandpass filtered by the analog frontend at a center frequency of 15.625 MHz and sampled at 4 samples per wavelength at this new center frequency. This frequency was chosen due to a maximum sampling frequency constraint of 62.5 MHz on the Vantage's analog to digital converter.

A real-time display mode was used to select an imaging plane, and then high-speed plane-wave acquisition was initiated. After all of the transmissions, pre-beamformed channel data were streamed to the host PC and the system was returned to the real-time mode. Beamformed in-phase (I) and quadrature (Q) signal components were generated using a graphic-processing-unit (GPU) based algorithm with an unsteered, 128-element Hann-apodized receive aperture [16].

**Table 1.** The relevant transducer specifications and system settings are given below.

| Parameter | Value |
|---|---|
| Excitation Frequency | 17.857 MHz |
| Received Center Frequency | 15.625 MHz |
| Received Center Wavelength | 98.56 μm |
| Received Bandwidth | 14–22 MHz |
| Elevation Focus | 8 mm |
| Element Height | 1.5 mm |
| Element Width | 80 μm |
| Element Pitch | 100 μm/1.01λ |

#### 2.2.2. Rotation Phantom

In order to assess the performance of the transmit sequences, an experiment was conducted using a rotating cylindrical hyperechoic phantom (ATS Laboratories, Bridgeport, CT, USA). The 10-mm diameter phantom was mounted onto a motor shaft and aligned with the linear array transducer (Figure 3). The phantom and transducer were submerged in a degassed water bath and a piece of sound absorbing material (Sorbothane Inc., Kent, OH, USA) was placed behind the phantom to suppress reverberation artifacts. The rotation phantom was spun at a constant rate of 12 revolutions per second. Data were acquired for two full revolutions of the phantom and beamformed. A position-based external trigger was used to ensure that all acquired data would be temporally synchronized.

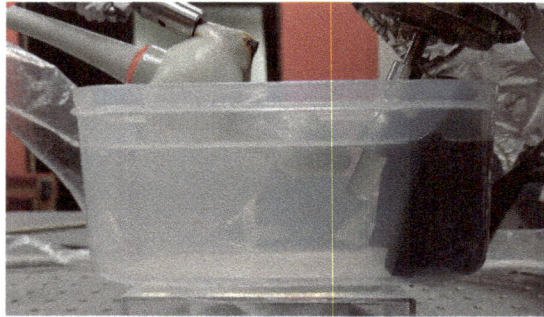

**Figure 3.** Experimental Setup. A linear 128-element linear array transducer was submerged in a degassed water bath and aligned with a hyperechoic rotation phantom. A layer of sound absorbing material (Sorbothane Inc., Kent, OH, USA) was used to suppress reverberation echoes from the edge of the plastic container.

#### 2.2.3. Transmission Sequences

Two of the transmission sequences were standard linear three-angle single transmit sequences (sTx3, $\{-5, 0, 5\}$) or seven-angle single transmit sequences (sTx7, $\{-5 : 5/3 : 5\}$). The third sequence was a double transmit sequence using 3 angles and, therefore, was composed of 6 transmits (dTx6, $\{-5, -5, 0, 0, 5, 5\}$). In each case, an odd number of angles was chosen to include the 0-degree steered transmit, which is least susceptible to grating lobes. Under this constraint, the sTx7 dataset was chosen to be a partial control for the sequence duration.

Each dataset was processed using the LRIs and HRIs using the lag-one estimator described in Section 2.1.5. For the HRIs, a standard sixth order infinite impulse response filter (IIR) Butterworth filter was used for clutter filtering, with a slow-time cutoff frequency of 500 Hz ($\approx$24.6 mm/s). The cutoff frequency of this filter was chosen to compensate for the slow transition band of the Butterworth filter, which takes approximately 300 Hz to reach a 60-dB attenuation level. For the LRIs, clutter filtering was performed on the matched angle LRI channels as described in Section 2.1.4, with similarly designed specifications at the appropriately reduced sample rates as shown in Figure 4. This design decision resulted in a periodic notching effect, in accordance with multirate signal processing theory [14,15].

The expected Nyquist velocities for each of these approaches are given in Table 2, in addition to other relevant slow-time quantities. In addition to the traditional approaches described above, the dTx6 dataset also was post processed using the pairwise (PW) methods described in Sections 2.1.6 (dTx6 LRI PW) and 2.1.7 (dTx6 HRI PW). The baseline averaging interval for each processing scheme was chosen to be around 1 ms in each case. Because the frame rate of the autocorrelation sequence varied so much across all the different processing schemes, matching the averaging schemes exactly was not possible. Furthermore, to avoid introducing half sample lags, odd length averages were used to keep the sequences temporally synchronized. These decisions resulted in the slight variations in the averaging interval as shown in Table 2.

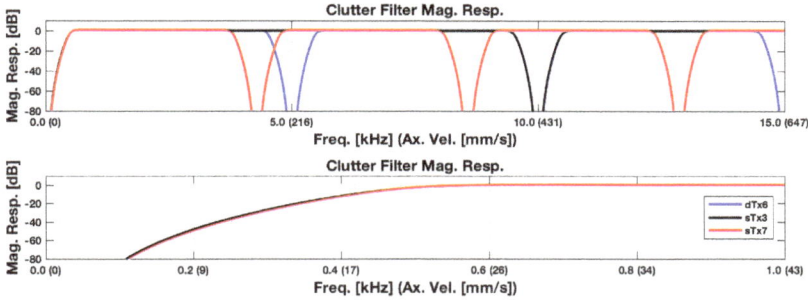

**Figure 4.** Clutter filter magnitude response. One should note the induced periodicity of the stop bands. This periodicity is due to the design decision to filter the LRI polyphase channels independently, as explained in Section 2.1.4. The repeated notch locations of the filter are located at integer multiples of $2N_T/f_N$. These notch locations correspond to the frame rates of HRIs, and consequentially the individual LRI polyphase channel frame rates as well. A zoomed in version is displayed below to help visualize stop band performance.

In addition to the color flow Doppler images, power Doppler images were also generated to help visualize the spatial energy distributions. These images were generated by calculating the mean-square value of the slow-time signal over the entire slow-time dimension, which was then converted to a decibel scale, such that

$$P(\vec{r}) = 10\log_{10}\left(\frac{1}{N}\sum_{n=0}^{N-1} s(\vec{r}, n\Delta t)^2\right), \tag{22}$$

where $P(\vec{r})$ is the power Doppler signal at the location $\vec{r}$, $N$ is the total number of slow-time samples, and $s(\vec{r}, t)$ is the slow-time signal for either the LRIs or HRIs as appropriate. Note that in the LRI case, we do not evaluate the power Doppler on the individual polyphase channels, but rather the full slow-time signal. For the pairwise estimation schemes, no power Doppler images were produced due to lack of a relevant metric which accounts for the pairwise nature of the signals. In practice, the LRI power Doppler could serve as a relevant metric for the pairwise approaches for the purposes of stationary clutter suppression.

As a ground truth for comparison purposes, a rotation model was fit to the data. Using a B-Mode image as a reference, points along the phantom boundaries were manually selected, and the centroid was computed. A 10-mm circular mask was then centered at the estimated centroid, and the axial component of the velocity was modeled using

$$v_z(x) = \omega_{\text{Phantom}} \times (x - \bar{x}) \tag{23}$$

where $\omega_{\text{Phantom}}$ is the angular velocity of the phantom and $\bar{x}$ is the $x$-component of the centroid. Using this ground truth model, the root-mean-square (RMSE) and bias errors were quantified. Because the phantom was also prone to precession, the standard deviation was also computed as a reference metric to account for any precessional and alignment biases. To facilitate easy comparison between all the methods, each metric was averaged over slow-time and depth so that they could be overlaid on the same plot for comparison purposes.

**Table 2.** The relevant slow-time quantities are given below. Note that in the case of the pairwise (PW) estimates in the last two rows the correlation frame rate differs from the in-phase and quadrature (IQ) frame rate. This is a characteristic of the pairwise estimation schemes. sTx3: three-angle single transmit sequence; sTx7: seven-angle single transmit sequence. dTx6: three-angle, double transmit sequence.

| Estimate | IQ Frame Rate (kHz) | Corr. Lag (μs) | Nyquist Velocity (mm/s) | Corr. Frame Rate (kHz) | Averaging Interval (Samples) | Averaging Interval (ms) |
|---|---|---|---|---|---|---|
| sTx3 HRI | 10 | 100.0 | 246.4 | 10 | 11 | 1.100 |
| sTx7 HRI | 4.286 | 233.3 | 105.6 | 4.286 | 5 | 1.167 |
| dTx6 HRI | 5 | 200.0 | 123.2 | 5 | 5 | 1.000 |
| sTx3 LRI | 30 | 33.3 | 739.2 | 30 | 31 | 1.033 |
| sTx7 LRI | 30 | 33.3 | 739.2 | 30 | 31 | 1.033 |
| dTx6 LRI | 30 | 33.3 | 739.2 | 30 | 31 | 1.033 |
| dTx6 LRI PW | 30 | 33.3 | 739.2 | 15 | 15 | 1.000 |
| dTx6 HRI PW | 30 | 33.3 | 739.2 | 5 | 5 | 1.000 |

## 3. Results

The results of the rotation phantom experiment are shown in Figures A1–A3. As can be observed in Figure A1, all the HRI estimates exhibit some degree of aliasing, as evidenced by the large errors at high velocities, with the three-angle single transmit sequence (sTx3 HRI) exhibiting the widest usable velocity range of all processing schemes using traditional processing. This result is predicted by the traditional Doppler slow-time frequency analysis, because with only three transmit events, the compounded dataset has the highest slow-time frame rate of the group, and thus the highest velocity range. Note that in each case, the aliasing error begins at lower velocities than predicted in Table 2. This is due to the broadband nature of the excitation. Because the excitation possesses frequency content above the center frequency, aliasing begins before the prediction, because the high-frequency components alias before the center frequency reaches its Nyquist rate. More conservative estimates could have been done by considering the entire band of the excitation, but such estimation schemes deviate from established conventions and were thus avoided to facilitate easy comparison.

Comparing the traditional lag-one estimates of the LRI processing schemes, one can observe that the sTx3 dataset is characterized by large RMSE and standard deviations across the velocity range. This would suggest low correlations in the individual slow-time lines. It is possible that in this case the sidelobe motion might have been fast enough to corrupt the estimate due to the high angular step, even despite the aggressive clutter filtering employed. This could possibly explain why the performance is recovered in the sTx3 HRI scheme.

For the seven-angle dataset (sTx7), the slow-time signal error resonates at around 200 mm/s. This error is anticipated due to the clutter filter stop band at that particular velocity range (see Figure 4). The effect also manifests in the power Doppler image in this dataset (Figure A5), because the signal is attenuated in the corresponding spatial region. This error is also exacerbated due to the sidelobe motion being present at those frequencies. While one could argue that such estimates would be suppressed by power Doppler thresholding, the fact remains that there is a weak spot in the velocity range that cannot be overcome using a conventional transmit sequence. Similarly, the velocity estimate is also extremely poor near the stationary regime, but because the signal-to-noise ratio (SNR) is low here, this is to be expected.

This secondary error resonance at the compounded frame rate is not present using the traditional double-transmit method (dTx6). Even though the signal power is lower at these velocities, the fact that the sidelobes move at a slower rate allows for a more robust estimate across a larger range than the sTx7 dataset, despite having comparable channel frame rates. The error in the near stationary regime is also markedly better than both the sTx LRI processing schemes. Both pairwise estimation schemes had only minor improvements in the dTx dataset, suggesting that the dominant improvement is the temporal non-uniformity of the sidelobe motion. The interleaved compounded dataset (dTx6 HRI PW), offered a compromise between full compounding (dTx6 HRI) and no compounding (dTx6 LRI), as it tracks well in low velocity regime, but not quite as well in the higher velocity regime.

## 4. Conclusions

In this study, we investigated the effects of using a double transmit sequence to simultaneously expand the Nyquist velocity range in such a way that it is resistant to sidelobe motion. Use of the double transmit sequence allowed for significant performance improvements in velocity regimes that would typically be attenuated by clutter filtering, demonstrating increased robustness to noise. The proposed methods significantly outperform the traditional methods in the high velocity regime in both root-mean square-error and standard deviation, with minimal performance losses in the low-velocity regime. These results suggest a new paradigm in velocity estimation in the context of high-frame-rate Doppler ultrasound.

**Acknowledgments:** This research was supported by grants from the National Institutes of Health (EY025215, EB022950, and R21EB020766).

**Author Contributions:** J.A.K. conceived the project. A.S.P. and J.A.K. proposed the new methods, and also designed and carried out the experiments. A.S.P. and M.L.O. analyzed the method and the results. A.S.P. drafted the manuscript; J.A.K. and M.L.O. revised and edited the manuscript. All authors read and approved the manuscript.

**Conflicts of Interest:** The authors declare no conflict of interest.

## Abbreviations

The following abbreviations are used in this manuscript:

PRF     Pulse repetition frequency
HRI     High-resolution image/compounded image
LRI     Low-resolution image/pre-compounded image
PSF     Point spread function
RMSE    Root-mean-square error
sTx     Single transmit sequence (dataset prefix)
dTx     Double transmit sequence (dataset prefix)
PW      Pairwise, (estimation scheme suffix)
SNR     Signal-to-noise ratio

## Appendix A. Multirate Theorems of Coherent Plane Wave Compounding

Let $s[n] \equiv s_a(\vec{r}, n\Delta t)$ be the discrete, slow-time, precompounded, analytic signal at a location $\vec{r}$. For compactness of notation, the position argument $\vec{r}$ is suppressed in the following section, with the understanding that it is evaluated independently at every spatial location.

**Theorem A1.** *The coherent plane wave compounded signal $s_+[n]$ is equivalent to a rectangular filtering of $s[n]$ followed by a downsampling operation.*

**Proof.** By definition, $s_+[n]$ is given by

$$s_+[n] \triangleq \sum_{n'=0}^{N_T-1} s\left[N_T n + n'\right]. \tag{A1}$$

Let us define the rectangular signal that is nonzero on the closed interval $[a, b]$ as

$$r_{[a,b]}[n] \equiv \begin{cases} 1, & n \in [a, b] \\ 0, & \text{otherwise.} \end{cases} \tag{A2}$$

A simple variable substitution of $n' \rightarrow -n'$ results in

$$s_+[n] = \sum_{n'=-(N_T-1)}^{0} s\left[N_T n - n'\right] \tag{A3}$$

$$= \sum_{n'=(1-N_T)}^{0} s\left[N_T n - n'\right] \tag{A4}$$

$$= \sum_{n'=-\infty}^{\infty} r_{[1-N_T,0]}[n']s\left[N_T n - n'\right] \tag{A5}$$

$$= s_f\left[N_T n\right], \tag{A6}$$

where $s_f[n]$ is the rectangularly filtered analytic signal given by

$$s_f[n] = r_{[1-N_T,0]}[n] \underset{n}{*} s[n] \tag{A7}$$

$$= \sum_{n'=-\infty}^{\infty} r_{[1-N_T,0]}[n']s\left[n - n'\right]. \tag{A8}$$

Because the downsampling operation by a factor of $N_T$ is defined as

$$\mathscr{D}_{N_T}\left\{s_f[n]\right\} = s_f[N_T n] \tag{A9}$$

we find that

$$s_+[n] = \mathscr{D}_{N_T}\left\{s_f[n]\right\} \tag{A10}$$

$$= \mathscr{D}_{N_T}\left\{r_{[1-N_T,0]}[n] \underset{n}{*} s[n]\right\}. \tag{A11}$$

Thus, the coherent compounding operation is equivalent to a rectangular filter of length $N_T$ followed by a downsampling of $N_T$. $\square$

**Theorem A2.** *Consider the traditional pipeline of passing the compounded signal $s_+[n]$ through a high-pass clutter filter with impulse response $h[n]$. Then, the output signal $s_c[n]$ is equivalent to filtering each individual polyphase channel $s_{\phi_{n'}}[n]$ and summing.*

**Proof.** Let us denote the output of the clutter filter as $s_c[n]$, such that

$$s_c[n] = h[n] \underset{n}{*} s_+[n]. \tag{A12}$$

Thus, we have

$$s_c[n] = h[n] \underset{n}{*} \left(\sum_{n'=0}^{N_T-1} s[N_T n + n']\right) \tag{A13}$$

Let us define the polyphase channel $s_{\phi_{n'}}[n]$ as the slow-time signal associated with the $n'$th transmit in the transmit sequence period. Mathematically the relationship is described by

$$s_{\phi_{n'}}[n] \equiv s[N_T n + n']. \tag{A14}$$

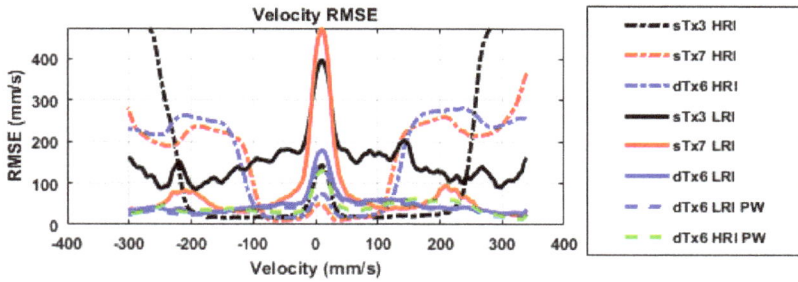

**Figure A1.** In the HRI estimators, the aliasing manifests as plateauing in the root-mean-square error (RMSE), resulting from the lower frame rate. All other estimators appear to predominately track the standard deviation shown in Figure A2.

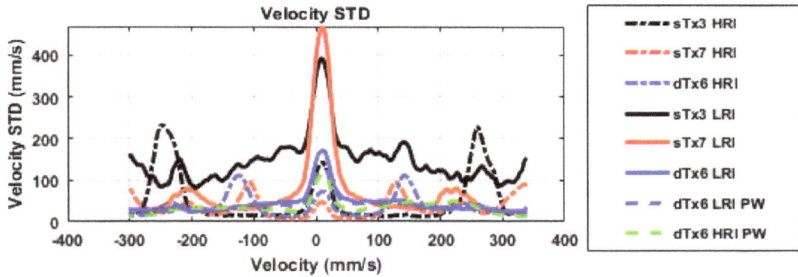

**Figure A2.** For the traditional HRI estimators, error resonates at the Nyquist velocities, likely due to the aliasing. The sTx3 LRI estimator appears to have uncharacteristically high error across the velocity range, suggesting that the individual LRIs are highly decorrelated, possibly due to a high angular step. In the sTx7 LRI dataset, the error resonates at the velocity associated with the transmit sequence period, likely due a combination of sidelobe motion and attenuation due to the clutter filter. The dTx6 estimators do not have a corresponding peak, suggesting an inherent robustness to these effects.

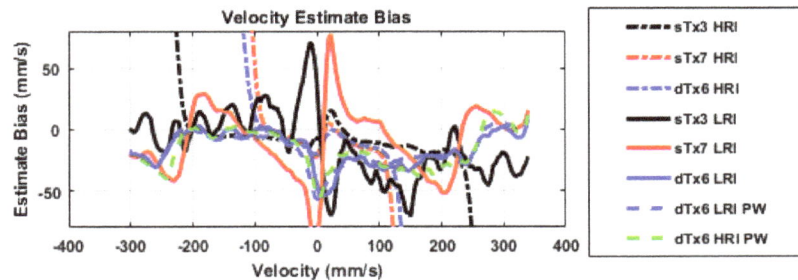

**Figure A3.** All estimators exhibit a slight negatively sloping trend, suggesting a possible alignment error in the ground truth. The HRI estimators each show evidence of aliasing at the predicted Nyquist velocities.

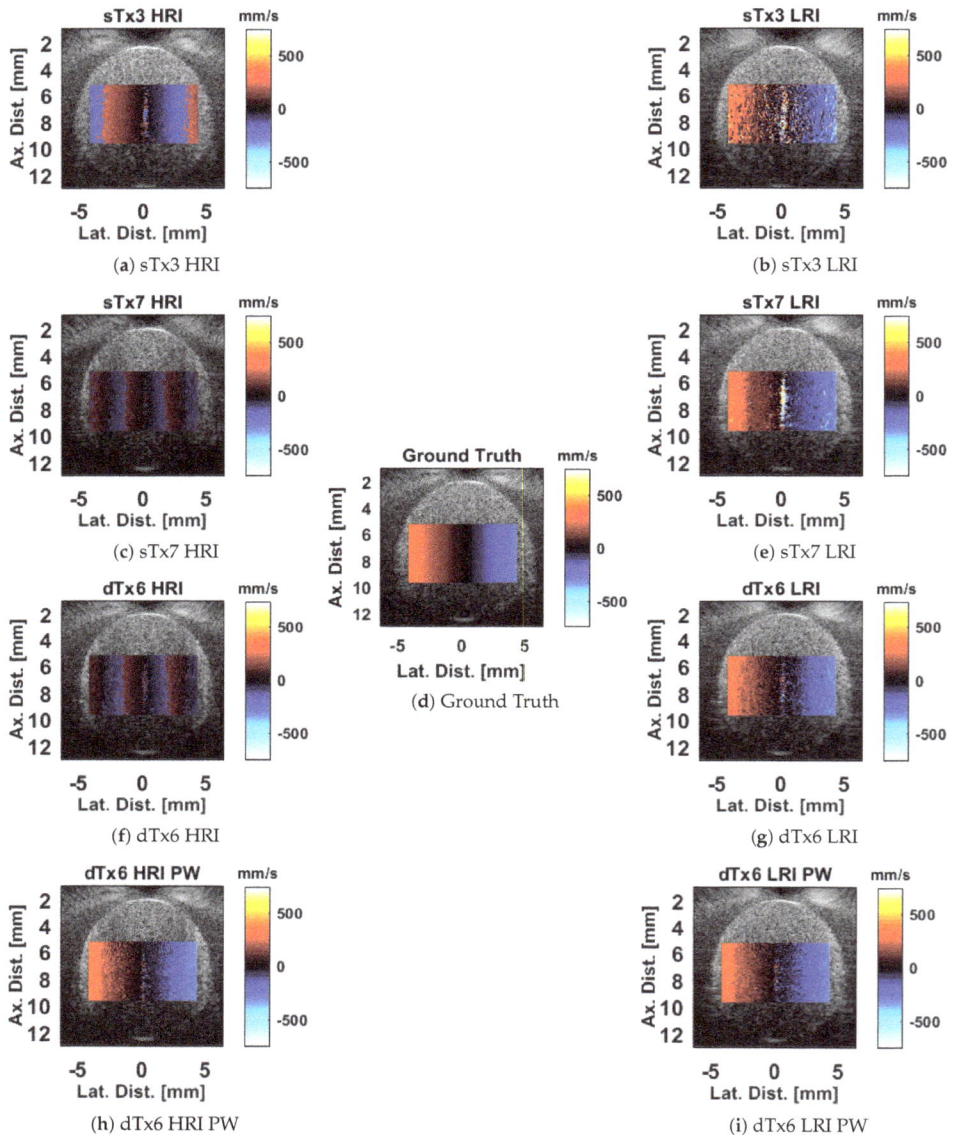

**Figure A4.** Color flow Doppler images generated from the experimental data. The HRI images are shown on the left, and the LRI images are shown on the right, with each row corresponding to a different dataset. The pairwise results are shown in the last row. All images are rendered with the same color scale to facilitate easy comparisons. Aliasing is clearly observed in the traditional HRI cases, manifesting as banding artifacts. The high error observed in the center is due to the clutter filter, which attenuates slow-moving components of the signal. This confirmed in the power Doppler images (Figure A5), which exhibit deep notches in the center of the image, corresponding to stationary axial motion.

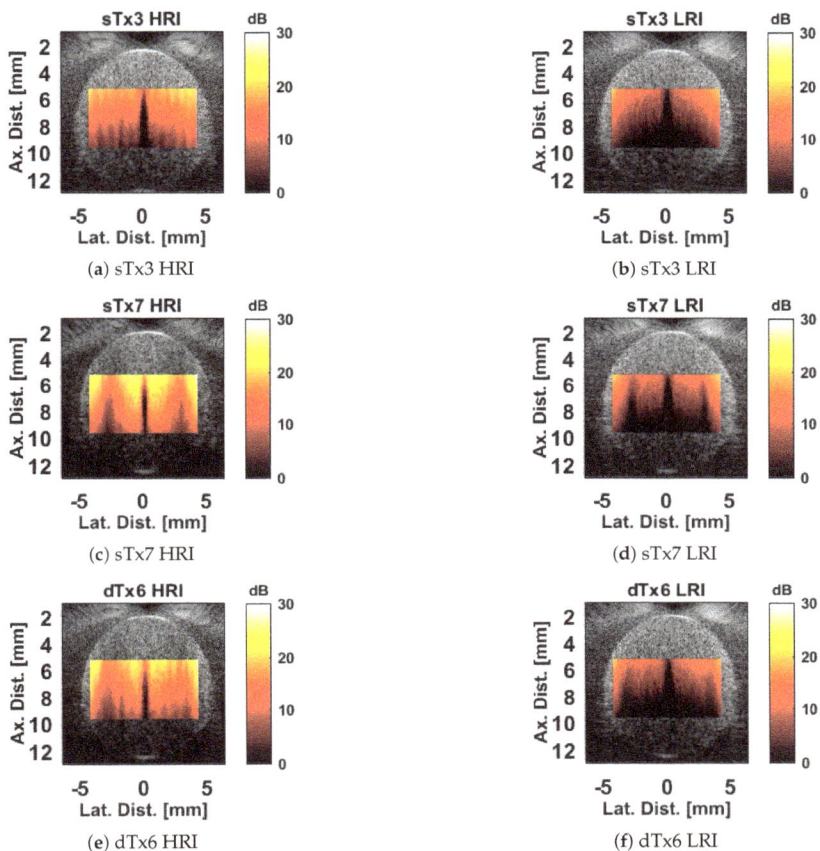

**Figure A5.** Power Doppler images generated from the experimental data. The HRI images are shown on the left, and the LRI images are shown on the right, with each row corresponding to a different dataset. The pairwise results are omitted, since they lacked a descriptive metric to characterize the power in such a way that would account for the pairwise estimation scheme. One should note the presence of spatial notches in the images. These notches correspond to velocities in the clutter filter stop band. The notches appear to spread out with depth, which is believed to be due the cumulative effect of attenuation. All images are rendered with the same color scale to facilitate easy comparisons. The lateral banding the sTx3 and dTx6 HRI processing schemes is believed to be due to an artifact in the broadside transmission that is specific to the Verasonics Vantage system. This artifact is likely due to the simultaneous excitation of all the elements. The effect is suppressed in the sTx7 dataset due to the additional steered excitations which do not exhibit this effect.

Thus, due to the commutativity of convolution and summation, we have

$$s_c[n] = h[n] \underset{n}{*} \left( \sum_{n'=0}^{N_T-1} s_{\phi_{n'}}[n] \right) \tag{A15}$$

$$= \sum_{n'=0}^{N_T-1} \left( h[n] \underset{n}{*} s_{\phi_{n'}}[n] \right). \tag{A16}$$

Therefore, filtering each individual polyphase component before summation yields the same response as filtering the compounded signal. $\square$

## References

1. Evans, D.H. Colour flow and motion imaging. *Proc. Inst. Mech. Eng. H* **2010**, *224*, 241–253.
2. Kasai, C.; Namekawa, K.; Koyano, A.; Omoto, R. Real-time two-dimensional blood flow imaging using an autocorrelation technique. *IEEE Trans. Son. Ultrason.* **1985**, *32*, 458–464.
3. Lu, J.Y. 2D and 3D high frame rate imaging with limited diffraction beams. *IEEE Trans. Ultrason. Ferroelectr. Freq. Contr.* **1997**, *44*, 839–856.
4. Lu, J.; Cheng, J.; Wang, J. High frame rate imaging system for limited diffraction array beam imaging with square-wave aperture weightings. *IEEE Trans. Ultrason. Ferroelectr. Freq. Contr.* **2006**, *53*, 1796–1812.
5. Cheng, J.; Lu, J.Y. Extended high-frame rate imaging method with limited-diffraction beams. *IEEE Trans. Ultrason. Ferroelectr. Freq. Contr.* **2006**, *53*, 880–899.
6. Montaldo, G.; Tanter, M.; Bercoff, J.; Benech, N.; Fink, M. Coherent plane-wave compounding for very high frame rate ultrasonography and transient elastography. *IEEE Trans. Ultrason. Ferroelectr. Freq. Contr.* **2009**, *56*, 489–506.
7. Hasegawa, H.; Kanai, H. High-frame-rate echocardiography using diverging transmit beams and parallel receive beamforming. *J. Med. Ultrason.* **2011**, *38*, 129–140.
8. Tong, L.; Gao, H.; Choi, H.F.; D'Hooge, J. Comparison of conventional parallel beamforming with plane wave and diverging wave imaging for cardiac applications: A simulation study. *IEEE Trans. Ultrason. Ferroelectr. Freq. Contr.* **2012**, *59*, 1654–1663.
9. Hartley, C.J.; Michael, L.H.; Entman, M.L. Noninvasive measurement of ascending aortic blood velocity in mice. *Am. J. Physiol.* **1995**, *268*, H499–505.
10. Wang, J.; Lu, J.Y. Motion artifacts of extended high frame rate imaging. *IEEE Trans. Ultrason. Ferroelectr. Freq. Contr.* **2007**, *54*, 1303–1315.
11. Denarie, B.; Tangen, T.A.; Ekroll, I.K.; Rolim, N.; Torp, H.; Bjastad, T.; Lovstakken, L. Coherent plane wave compounding for very high frame rate ultrasonography of rapidly moving targets. *IEEE Trans. Med. Imaging* **2013**, *32*, 1265–1276.
12. Poree, J.; Posada, D.; Hodzic, A.; Tournoux, F.; Cloutier, G.; Garcia, D. High-frame-rate echocardiography using coherent compounding with Doppler-based motion-compensation. *IEEE Trans. Med. Imaging* **2016**, *35*, 1647–1657.
13. Ekroll, I.K.; Voormolen, M.M.; Standal, O.K.V.; Rau, J.M.; Lovstakken, L. Coherent compounding in Doppler imaging. *IEEE Trans. Ultrason. Ferroelectr. Freq. Contr.* **2015**, *62*, 1634–1643.
14. Vaidyanathan, P. *Multirate Systems and Filter Banks*; Prentice Hall: Upper Saddle River, NJ, USA, 1993; pp. 100–108, 119, 120–122.
15. Proakis, J.G.; Manolakis, D.G. *Digital Signal Processing: Principles Algorithms and Applications*, 4 ed.; Pearson Education Inc.: Upper Saddle River, NJ, USA, 2007; pp. 654–727, 755–759, 760.
16. Yiu, B.Y.S.; Tsang, I.K.H.; Yu, A.C.H. GPU-based beamformer: fast realization of plane wave compounding and synthetic aperture imaging. *IEEE Trans. Ultrason. Ferroelectr. Freq. Contr.* **2011**, *58*, 1698–1705.

![applied sciences logo] *applied sciences*

MDPI

*Article*

# Iterative 2D Tissue Motion Tracking in Ultrafast Ultrasound Imaging

**John Albinsson [1], Hideyuki Hasegawa [2], Hiroki Takahashi [2], Enrico Boni [3], Alessandro Ramalli [3,4], Åsa Rydén Ahlgren [5,6] and Magnus Cinthio [1,\*]**

[1]  Department of Biomedical Engineering, Faculty of Engineering, Lund University, 221 00 Lund, Sweden; john.albinsson@bme.lth.se
[2]  Graduate School of Science and Engineering for Research, University of Toyama, Toyama 930-8555, Japan; hasegawa@eng.u-toyama.ac.jp (H.H.); takahashi.hiroki@xpost.plala.or.jp (H.T.)
[3]  Department of Information Engineering, University of Florence, 501 39 Florence, Italy; enrico.boni@unifi.it (E.B.); alessandro.ramalli@unifi.it (A.R.)
[4]  Laboratory of Cardiovascular Imaging and Dynamics, Department of Cardiovascular Sciences, KU Leuven, 3000 Leuven, Belgium
[5]  Department of Translational Medicine, Lund University, 221 00 Lund, Sweden; asa.ryden_ahlgren@med.lu.se
[6]  Department of Medical Imaging and Physiology, Skåne University Hospital, Lund University, 205 02 Malmö, Sweden
\*  Correspondence: magnus.cinthio@bme.lth.se; Tel.: +46-(0)46-222-9710

Received: 7 March 2018; Accepted: 21 April 2018; Published: 25 April 2018

**Abstract:** In order to study longitudinal movement and intramural shearing of the arterial wall with a Lagrangian viewpoint using ultrafast ultrasound imaging, a new tracking scheme is required. We propose the use of an iterative tracking scheme based on temporary down-sampling of the frame-rate, anteroposterior tracking, and unbiased block-matching using two kernels per position estimate. The tracking scheme was evaluated on phantom B-mode cine loops and considered both velocity and displacement for a range of down-sampling factors ($k = 1$–$128$) at the start of the iterations. The cine loops had a frame rate of 1300–1500 Hz and were beamformed using delay-and-sum. The evaluation on phantom showed that both the mean estimation errors and the standard deviations decreased with an increasing initial down-sampling factor, while they increased with an increased velocity or larger pitch. A limited in vivo study shows that the major pattern of movement corresponds well with state-of-the-art low frame rate motion estimates, indicating that the proposed tracking scheme could enable the study of longitudinal movement of the intima–media complex using ultrafast ultrasound imaging, and is one step towards estimating the propagation velocity of the longitudinal movement of the arterial wall.

**Keywords:** ultrafast ultrasound imaging; block-matching; speckle tracking; arterial longitudinal wall movement; in vivo

## 1. Introduction

Ultrafast ultrasound imaging has been used as the basis for the development of a number of methods intended for diagnosing and exploring different phenomena in vivo, e.g., shear wave elastography [1–4], acoustic radiation force impulse imaging [5], vector flow imaging [6–8], and a method for skeletal muscle contraction [9], functional ultrasound imaging of the brain [10], and cardiac motion [11,12]. The arterial walls have been investigated by estimating the radial strain in the common carotid artery [13] and the radial pulse wave velocity [14,15].

In cardiovascular research, the radial movement of the arterial wall, i.e., the diameter change, has been the subject of extensive research, forming the basis for estimation of arterial wall stiffness [16].

Increased stiffness of the large central arteries has been shown to be an independent risk factor for cardiovascular mortality [17]. In contrast to the radial movement, the longitudinal movement of the arterial wall has gained less attention. We have, however, shown that in both large predominantly elastic arteries and in large muscular arteries there is a distinct bi-directional displacement of the arterial wall during the cardiac cycle [18]. The intima–media of these arteries exhibits a longitudinal displacement that is larger than that of the adventitial region [18] and thus, there is shear strain and shear stress within the arterial wall [18–21]. We have recently reported that longitudinal movement and intramural shear strain undergo profound changes in response to the important circulatory hormones adrenalin and noradrenalin [22], indicating that the longitudinal movements and resulting intramural shear strain can constitute an important but overlooked mechanism in the cardiovascular system. Studies have indicated that the maximal amplitude of the longitudinal displacement of the common carotid artery is reduced in subjects with cardiovascular risk factors [23], and suspected and manifest atherosclerotic disease [24,25]. However, the physiology behind the observed longitudinal vessel wall movement pattern is largely unknown.

It is our belief that the use of ultrafast ultrasound imaging in combination with 2D tissue motion estimation can increase our understanding of this phenomenon and make it possible to estimate the propagation velocity of the longitudinal movement. However, to explore the longitudinal movement of the arterial wall by using ultrasound, the artery is scanned in the longitudinal direction and the longitudinal movement of the arterial wall occurs in the lateral direction of the ultrasound image. It is problematic to estimate lateral tissue motion in ultrafast ultrasound imaging in vivo as the tissue moves only a very short distance between consecutive frames due to the high frame rate. Thus, the motion to be estimated will be very small compared to the expected uncertainty in the motion estimates caused by the limited signal-to-noise ratio. The estimation uncertainty is larger in the lateral direction [26] as ultrasound frames normally have lower spatial resolution in the lateral direction. Consequently, Lagrangian tracking in the lateral direction and in every frame is very likely to give a large accumulated error even with an unbiased motion estimator when using ultrafast ultrasound imaging. The motion estimations can be improved by averaging motion estimations over multiple frames, but this will decrease the effective frame rate and will function as a low-pass filter on the motion estimations in the time domain. This can potentially hide vital information in the motion estimations.

In this paper, we propose to estimate 2D motions with a Lagrangian viewpoint in ultrafast ultrasound cine loops using an iterative motion estimation tracking scheme in which the initial length between the used frames is larger than one. Contrary to phase-sensitive motion estimation methods (e.g., [14,27]) where the estimated motion must be small to avoid aliasing, our experience shows that the relative motion estimation error decreases for block-matching methods when the length of the estimated motion increases [26]. Since the motion between two frames in ultrafast ultrasound cine loops is often very small and the speckle decorrelation is limited, the risk for the speckle decorrelation over several, e.g., 128, frames is small but the total motion over this number of frames will be larger and easier to accurately estimate using block-matching. Therefore, we propose a temporary down sampling of the frame rate in which a first Lagrangian motion estimation is performed between each $k$ frame, e.g., initial frame interval $k = 128$. The cine loop is thereafter iteratively re-sampled with shorter frame intervals and the position of the kernel in one in-between frame can be estimated using the two kernels of the anteroposterior frames as reference kernels. The tracking scheme is hypothesized to reduce the size of the accumulative errors both by using two separate motion estimations for each estimated position, thus reducing each estimation error, and by using much fewer estimations from the start of the tracking before reaching the investigated frame.

The aim of this study was to evaluate the proposed 2D tissue motion estimation tracking scheme in ultrafast ultrasound cine loops. In a phantom evaluation, the proposed tracking scheme was evaluated for a range of initial down-sampling factors ($k = 1$–$128$). The motion estimation errors of the proposed tracking scheme using ultrafast ultrasound cine loops were compared to those obtained in low frame rate cine loops, obtained with conventional beamforming. The motion estimation errors of both velocity and displacement were evaluated. The tracking performance was evaluated using a 100 μm pitch transducer and a 200 μm pitch transducer. The feasibility of using the proposed 2D tissue motion estimation tracking scheme in vivo was evaluated in a limited in vivo study.

## 2. Materials and Methods

The in-house block-matching method, developed to estimate the location of the target in a given frame, builds on works by Albinsson et al. [26,28] and will be summarized below. Here, we propose a novel tracking scheme that is based on the re-sampling of the cine loop along the time axis. The tracking performance of the novel 2D tissue motion estimator was evaluated on phantom and in vivo cine loops. Furthermore, the tracking performance obtained in ultrafast cine loops was compared to that achieved in low frame rate cine loops, obtained with conventional beamforming.

### 2.1. Proposed Tracking Scheme

The method denoted as the "basic method" in [28] is a sparse iterative block-matching method, that uses the sum of absolute differences as the matching criterion and an unlimited search area. In this work, the sub-sample method has been replaced with the method presented below. The method denoted "basic method using an extra reference block" in [28] uses two independent kernels from two consecutive frames. The search area of the second kernel is limited to a small area around a position determined by the "basic method".

The proposed motion estimation tracking scheme consists of two parts:

1. First, the frame rate is temporarily down sampled by a factor $k$, where $k = (2, 4, 8, 16, 32, 64,$ or $128)$. The position of each kernel is estimated with a Lagrangian viewpoint between every frame in the temporary cine loop (solid lines in Figure 1a). The position of the kernel in each frame is estimated using a block-matching method with an extra kernel described in [28] (where the method is denoted as the "basic method using an extra reference block"). This method was developed to minimize estimation errors when using a Lagrangian viewpoint.

2. Iteratively: the frame rate is temporarily down sampled by a factor $m = k/2^i$ where $i$ is the iteration number. The unknown kernel positions in each middle frame in the temporary cine loop are determined by the kernels from the anteroposterior frames (dashed lines in Figure 1b). The two independently estimated positions are averaged to determine the kernel position in the middle frame. The iterations continued until the position of the kernel is estimated in every frame (dashed lines in Figure 2c), i.e., $m = 1$. The position of the kernel is estimated using the block-matching method denoted "basic method" in [28].

The sub-sample estimation in both part 1 and 2 is first performed by parabolic interpolation; if the estimate is $y \pm 0.15$ pixels, where $y$ is any natural number, the estimate is used, otherwise a modified grid slope sub-sample estimator is used to recalculate the estimate [26]. Parabolic and grid slope interpolation complement each other: parabolic interpolation is biased for sub-sample estimation close to $y \pm 0.5$ pixels, whereas grid slope interpolation gives noisy estimates close to $y \pm 0.0$ pixels.

The size of a kernel in the phantom study was 1 mm axially and laterally. This resulted in an axial kernel size of 41 pixels, and a lateral kernel size of 11 pixels using 100 μm line distance and 5 pixels using 200 μm line distance. The size of a kernel in the in vivo study was 0.6 mm axially and 3.8 mm laterally.

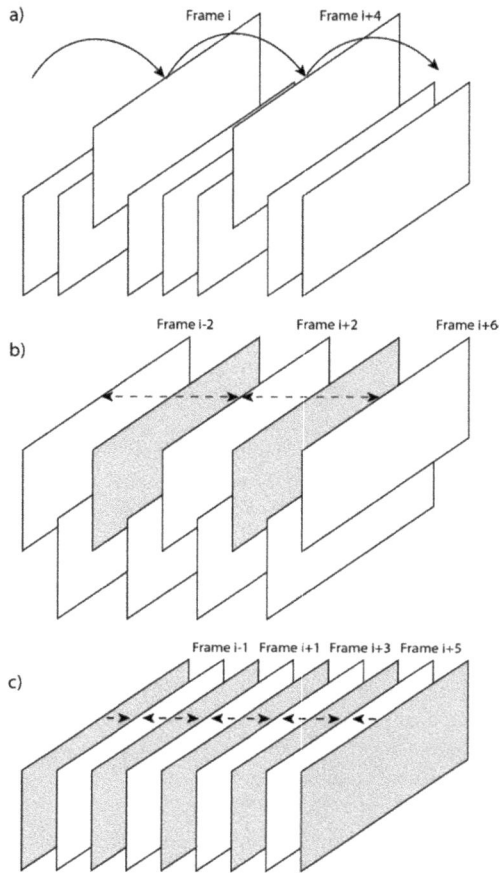

**Figure 1.** The figure shows which frames to use in (**a**) part 1 and (**b**,**c**) part 2 (1st and 2nd iteration) of the proposed tracking scheme for $k = 4$. The squares mark sampled frames and a raised frame marks a frame used in the current iteration. Please note that all frames are raised in (**c**) (2nd iteration). A gray frame marks a frame in which a position for the kernel has been estimated in a previous iteration. The base of the arrows shows the frame in which the kernel is collected and the point of the arrows shows in which frame the kernel is searched for. In (**b**,**c**) iteration, there are two arrows pointing at each frame in which to estimate the position. The estimated position in these frames is calculated as the average estimated position using two kernels from two different frames.

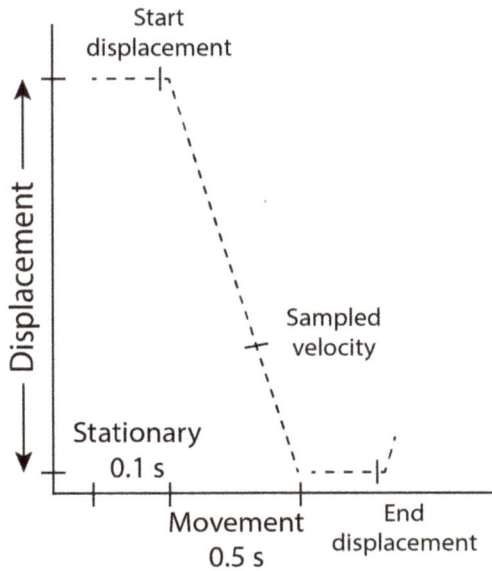

**Figure 2.** Scheme for the set motion of the transducer and the frames used for determining displacement and velocity. The displacement of each kernel was calculated as the difference between its position in the start and end frames. The velocity was estimated as the difference between two consecutive frames.

*2.2. Cine Loops*

The phantom cine loops were collected in Japan at a pulse repetition frequency of 5208 Hz by a 96-channel ultrasound scanner (RSYS0002, Microsonic, Tokyo, Japan) equipped with two linear array ultrasonic transducers with a center frequency of 7.5 MHz. The transducers had a pitch of 100 μm and 200 μm, respectively, which corresponds to $\lambda/2$ and $\lambda$. The pixel densities in the cine loops were 40.6 mm$^{-1}$ axially and either 10.0 mm$^{-1}$ or 5.0 mm$^{-1}$ laterally depending on the pitch of the transducer. Each transducer was moved diagonally repeatedly back and forth at constant velocity with short stops at each turning point (Figure 2) using automatic stages (ALS-6012-G0M and ALV-600H0M, Chuo Precision Industrial, Tokyo, Japan). A sponge was used to create realistic speckle. Two different velocities were used: 2.0 mm/s laterally and 1.0 mm/s axially with displacements of 1.0 mm laterally and 0.5 mm axially; and 1.0 mm/s laterally and 0.5 mm/s axially with displacements of 0.5 mm laterally and 0.25 mm axially. In the lateral direction, this corresponds to a displacement of 1.5 μm/frame and 0.77 μm/frame, respectively. The collected radio frequency data were beamformed using delay-and-sum [13]. In the present study, one frame was obtained from four plane wave transmissions resulting in a frame rate of 1302 Hz. A plane wave is transmitted with 96 active elements, and echo signals were received by the same elements. Each receiving beam was created using the echo signals obtained from 72 of 96 elements. Consequently, 24 receiving beams were created in one transmit event. Then, the active aperture was translated laterally by 24 elements, and the same procedure was repeated four times to obtain 96 receiving beams. In receive, a Hanning apodization was used. Cine loops were also collected at a lower frame rate (41 Hz) with a conventional linear scan scheme. These cine loops used the same transducers, transducer movements, and kernel size, while the motion estimations were conducted in every frame ($k = 1$, see above).

To evaluate the feasibility of using the proposed 2D tissue motion estimation tracking scheme in vivo, a limited in vivo study was conducted in Sweden. The in vivo experiment was performed using a 64-channel ULA-OP system [29,30] equipped with a 192-element LA435 linear array transducer

(Esaote SpA, Florence, Italy) with a 200 μm pitch. The cine loops were collected at 1500 Hz using a single plane wave transmission; the 64-line frames were beamformed using delay-and-sum with dynamic apodization having the f-number equal to 2. The line distance between the 64 lines was 200 μm. ECG was not available. To be able to compare with state-of-the-art estimation of the longitudinal movement of the arterial wall, low frame rate cineloops in vivo were collected using a Philips Epiq 7 (Philips Medical Systems, Bothell, WA, USA) equipped with a linear array transducer (model L18-5, Philips Medical Systems, Bothell, WA, USA). The right common carotid artery was scanned in the longitudinal direction, oriented horizontally in the image 2–3 cm proximal to the bifurcation. The healthy volunteers gave informed consent according to the Helsinki Declaration and the study was approved by the Ethics Committee, Lund University.

### 2.3. Evaluation of Motion Estimations

For each setting, two evaluation metrics for the motion estimation were calculated using 90 kernels distributed in six columns with no overlap laterally and 15 rows with 50% overlap axially. The first evaluation metric was the difference between the set displacement and the estimated displacement for each kernel. The displacements were calculated as the distance moved by each kernel between the start frame and the end frame (0.6 s after start frame) (Figure 2). The second evaluation metric was the difference between the set velocity and the estimated velocity of each kernel. The velocities were estimated as the motion between two consecutive frames 0.4 s after the start of tracking (Figure 2).

The statistical significance of changes in the mean estimation errors and standard deviations was tested for the initial length of iteration using $k = 1$ as the reference compared to other initial lengths of iteration and cine loops sampled at a low frame rate. The statistical significance of changes in the mean estimation errors was also tested between high frame rate cine loops using $k = 128$ as the initial length of iteration and cine loops sampled at a low frame rate. Significance testing was conducted with $p < 0.05$ as the significance level utilizing the Analysis of Variance (ANOVA) for changes in mean values and the two-sample $F$-test for changes in standard deviations. Because the ANOVA was balanced and the changes in the standard deviations were limited, the unequal standard deviations were deemed to have negligible influence on the tests.

The in vivo measurements were evaluated by visual comparisons of the plotted motion estimations. One comparison was performed between ultrafast ultrasound imaging and conventional ultrasound imaging. The second comparison was performed in the same cine loop using four different positions along the artery wall.

### 3. Results

Figure 3 shows an example of lateral tracking resulting from the proposed motion estimation method for $k = 1$, 16 and 128 in a high-frame cine loop. The tracking curves show that the normal frame-to-frame tracking ($k = 1$) drifted away while the proposed method tracked the movement better and better with increasing $k$.

Figures 4 and 5 show the lateral and axial estimation errors when estimating velocity. In general, the mean estimation errors and the standard deviation decreased with increased initial length of iteration (larger $k$). Increased velocity of the phantom increased the standard deviations. A smaller pitch decreased both the mean value and the standard deviation of the lateral estimation errors, while the axial estimation errors, for the most part, were unaffected. A smaller pitch was more important when using low frame rate imaging than when using high frame rate imaging.

Figures 6 and 7 show the lateral and axial estimation errors when estimating the displacement. In general, the mean estimation errors and the standard deviations decreased with increasing initial length of iteration. A smaller pitch decreased both the mean value and the standard deviation of the lateral estimation errors, while the axial estimation errors, for the most part, were unaffected.

Tables 1–4 present the estimation errors both when estimating velocity and when estimating the displacement. The lateral mean estimation error is often larger and the standard deviation is always larger in high frame rate cine loops using frame-to-frame tracking ($k = 1$) than in low frame rate cineloops ($p < 0.05$). The lateral mean estimation error is often smaller and the standard deviation is usually larger in high frame rate cine loops using iterative tracking ($k = 128$) than in low frame rate cineloops ($p < 0.05$).

Figure 8 shows the estimated movement of the intima–media complex of the common carotid artery wall of a 47-year-old healthy female using both ultrafast ultrasound imaging ($k = 64$) (solid lines) and conventional ultrasound imaging (dashed lines). The estimations in both cine loops clearly show a bi-directional longitudinal movement pattern of the same order of magnitude. The estimated movement curve, showing approximately three heartbeats, also indicates repeatability of the movement pattern. Figure 9 shows the estimated movement of the intima–media complex of the common carotid arterial wall of a 35-year-old healthy female at four different lateral positions along the vessel wall ($k = 32$).

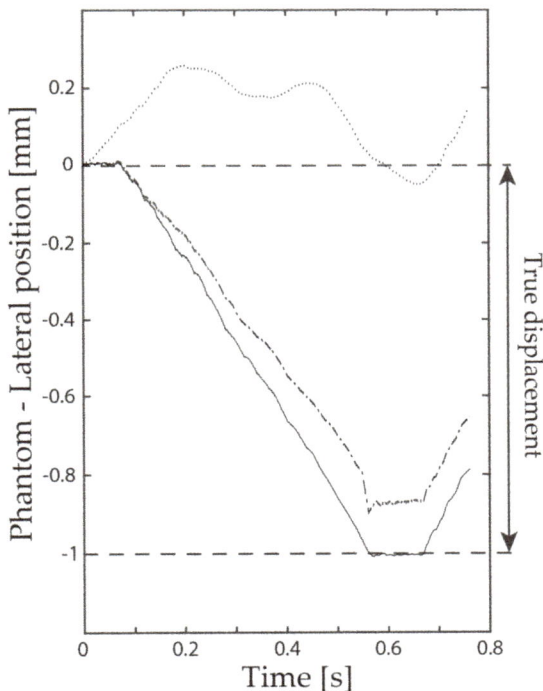

**Figure 3.** The lateral position of a kernel estimated by the proposed iterative tracking scheme with an initial length of iteration set to $k = 128$ (solid line), $k = 16$ (dash-dotted line), and $k = 1$ (dotted line). The 100 μm pitch transducer was displaced 1 mm laterally relative to the phantom with a lateral velocity of 2 mm/s while sampling 1302 frames per second.

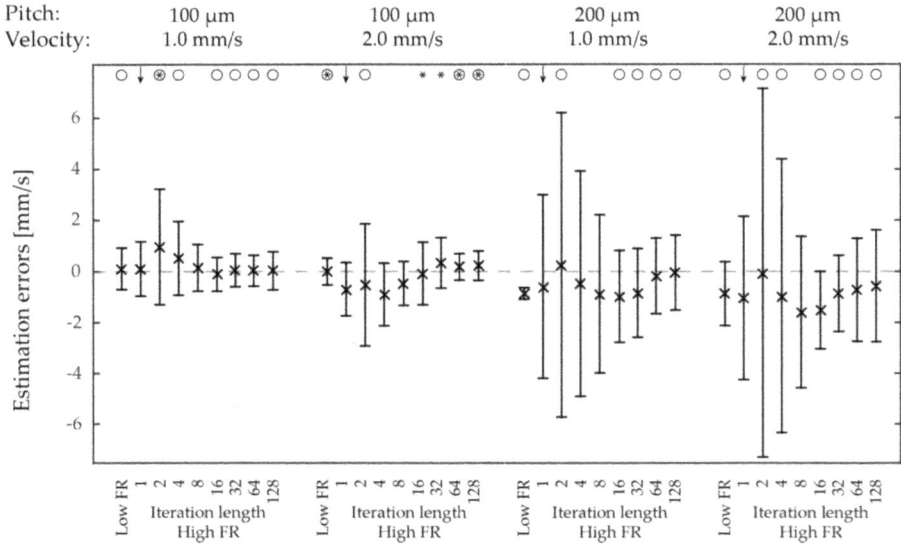

**Figure 4.** Lateral estimation errors of the velocity for two different velocities each using two different pitches (100 μm and 200 μm). The crosses indicate the mean estimation error with the error bars indicating ± one standard deviation. Significant changes ($p < 0.05$) were calculated compared to $k = 1$ in each setting (marked with an arrow). The star indicates a change in mean estimation error and a circle indicates a change in the standard deviation. FR signifies frame rate: low = 41 Hz, high = 1302 Hz.

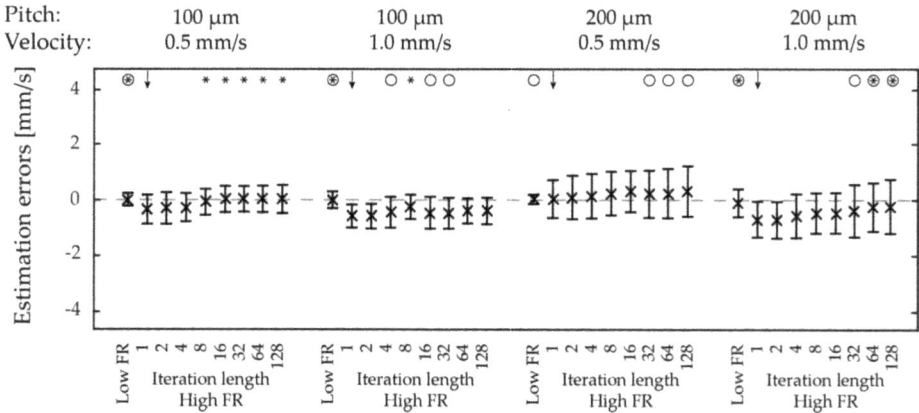

**Figure 5.** Axial estimation errors of the velocity for two different velocities each using two different pitches (100 μm and 200 μm). The crosses indicate the mean estimation with the error bars indicating ± one standard deviation. Significant changes ($p < 0.05$) were calculated compared to $k = 1$ in each setting (marked with an arrow). The star indicates a change in mean estimation error and a circle indicates a change in the standard deviation. FR signifies frame rate: low = 41 Hz, high = 1302 Hz.

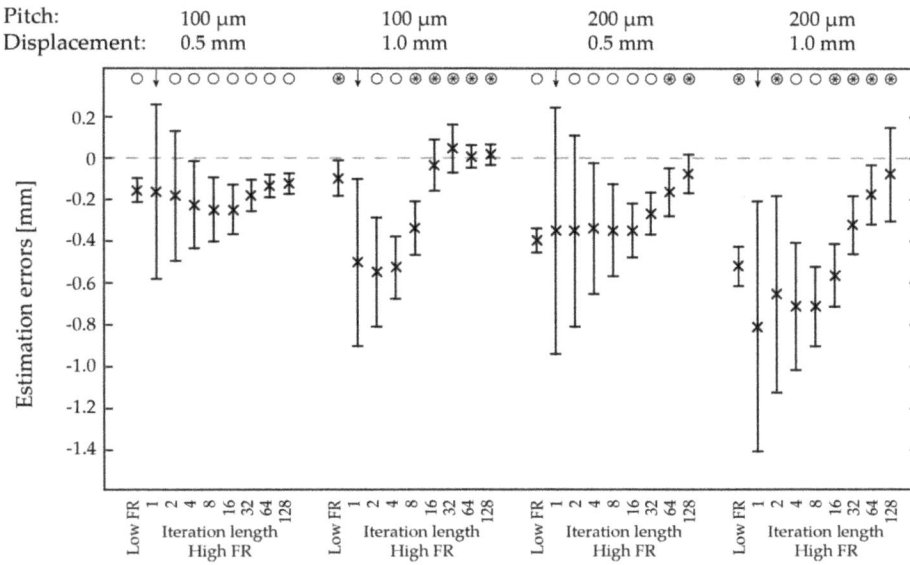

**Figure 6.** Lateral estimation errors of the displacement for two different displacements each using two different pitches (100 μm and 200 μm). The crosses indicate the mean estimation error for each setting with the error bars indicating $\pm$ one standard deviation. Significant changes ($p < 0.05$) were calculated compared to $k = 1$ in each setting (marked with an arrow). The star indicates a change in the mean estimation error and a circle indicates a change in the standard deviation. FR signifies frame rate: low = 41 Hz, high = 1302 Hz.

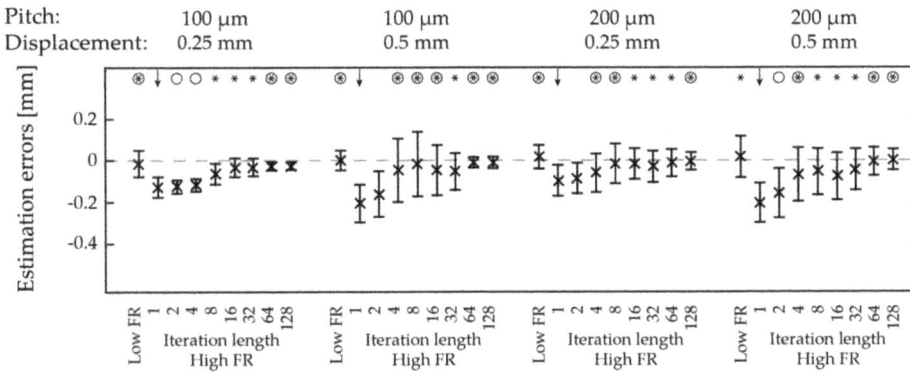

**Figure 7.** Axial estimation errors of the displacement for two different displacements each using two different pitches (100 μm and 200 μm). The crosses indicate the mean estimation error for each setting with the error bars indicating $\pm$ one standard deviation. Significant changes ($p < 0.05$) were calculated compared to $k = 1$ in each setting (marked with an arrow). The star indicates a change in the mean estimation error and a circle indicates a change in the standard deviation. FR signifies frame rate: low = 41 Hz, high = 1302 Hz.

**Table 1.** Lateral estimation errors of the velocity given as the mean estimation error ± one standard deviation.

| Pitch | | 100 μm | 200 μm | 100 μm | 200 μm |
|---|---|---|---|---|---|
| Velocity | | 1000 μm/s | 1000 μm/s | 2000 μm/s | 2000 μm/s |
| Significance | | - | a, b | - | - |
| Low FR | | 117 ± 805 | −852 ± 229 | 4 ± 520 | −871 ± 1253 |
| High FR | $k = 1$ | 108 ± 1058 | −598 ± 3595 | −691 ± 1044 | −1055 ± 3198 |
| | $k = 128$ | 32 ± 742 | −51 ± 1462 | 225 ± 570 | −572 ± 2184 |

All values are given in μm/s. Motion estimations were made using $k = 128$ for the high frame rate cine loops. Significance was defined as $p < 0.05$ in each column where a: Low frame rate (FR) vs. $k = 1$, b: Low frame rate (FR) vs. $k = 128$, and c: $k = 1$ vs. $k = 128$.

**Table 2.** Axial estimation errors of the velocity given as the mean estimation error ± one standard deviation.

| Pitch | | 100 μm | 200 μm | 100 μm | 200 μm |
|---|---|---|---|---|---|
| Velocity | | 500 μm/s | 500 μm/s | 1000 μm/s | 1000 μm/s |
| Significance | | a, c | a, b | - | a, c |
| Low FR | | 22 ± 228 | 39 ± 167 | 9 ± 302 | −90 ± 497 |
| High FR | $k = 1$ | −333 ± 502 | 50 ± 674 | −576 ± 415 | −681 ± 643 |
| | $k = 128$ | 35 ± 504 | 324 ± 896 | −381 ± 475 | −223 ± 967 |

All values are given in μm/s. Motion estimations were made using $k = 128$ for the high frame rate cine loops. Significance was defined as $p < 0.05$ in each column where a: Low frame rate (FR) vs. $k = 1$, b: Low frame rate (FR) vs. $k = 128$, and c: $k = 1$ vs. $k = 128$.

**Table 3.** Lateral estimation errors of the set displacement given as the mean estimation error ± one standard deviation.

| Pitch | | 100 μm | 200 μm | 100 μm | 200 μm |
|---|---|---|---|---|---|
| Displacement | | 500 μm | 500 μm | 1000 μm | 1000 μm |
| Significance | | - | a, b, c | b, c | a, b, c |
| Low FR | | −154 ± 58 | −397 ± 58 | −97 ± 86 | −521 ± 94 |
| High FR | $k = 1$ | −160 ± 421 | −347 ± 592 | −501 ± 400 | −807 ± 599 |
| | $k = 128$ | −123 ± 50 | −75 ± 93 | 16 ± 49 | −78 ± 226 |

All values are given in μm. Motion estimations were made using $k = 128$ for the high frame rate cine loops. Significance was defined as $p < 0.05$ in each column where a: Low frame rate (FR) vs. $k = 1$, b: Low frame rate (FR) vs. $k = 128$, and c: $k = 1$ vs. $k = 128$.

**Table 4.** Axial estimation errors of the set displacement given as the mean estimation error ± one standard deviation.

| Pitch | | 100 μm | 200 μm | 100 μm | 200 μm |
|---|---|---|---|---|---|
| Displacement | | 250 μm | 250 μm | 500 μm | 500 μm |
| Significance | | a, c | a, c | a, c | a, c |
| Low FR | | −15 ± 63 | 17 ± 57 | 1 ± 47 | 17 ± 100 |
| High FR | $k = 1$ | −127 ± 49 | −96 ± 74 | −206 ± 90 | −203 ± 94 |
| | $k = 128$ | −26 ± 18 | −4 ± 43 | −10 ± 28 | 3 ± 51 |

All values are given in μm. Motion estimations were made using $k = 128$ for the high frame rate cine loops. Significance was defined as $p < 0.05$ in each column where a: Low frame rate (FR) vs. $k = 1$, b: Low frame rate (FR) vs. $k = 128$, and c: $k = 1$ vs. $k = 128.4$.

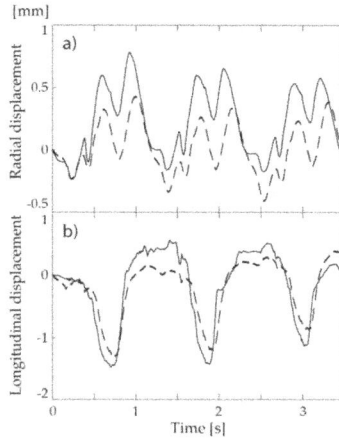

**Figure 8.** The estimated (**a**) radial and (**b**) lateral position of a kernel in vivo on the intima–media complex of the vessel wall of the common carotid artery in a healthy 47-year-old female obtained using ultrafast ultrasound imaging (solid line) and conventional ultrasound imaging (dashed line). The ultrafast ultrasound imaging was sampled at 1500 Hz and the proposed iterative tracking scheme had an initial length of iteration of $k = 64$ frames. The conventional ultrasound imaging was sampled at 99 Hz and the tracking was frame-by-frame. The main features of the in vivo curve estimated using a high frame rate cine loop agree well with our low frame rate in vivo measurements. No ECG signal was available for synchronizing the lines and the cine loops were collected 5 minutes apart.

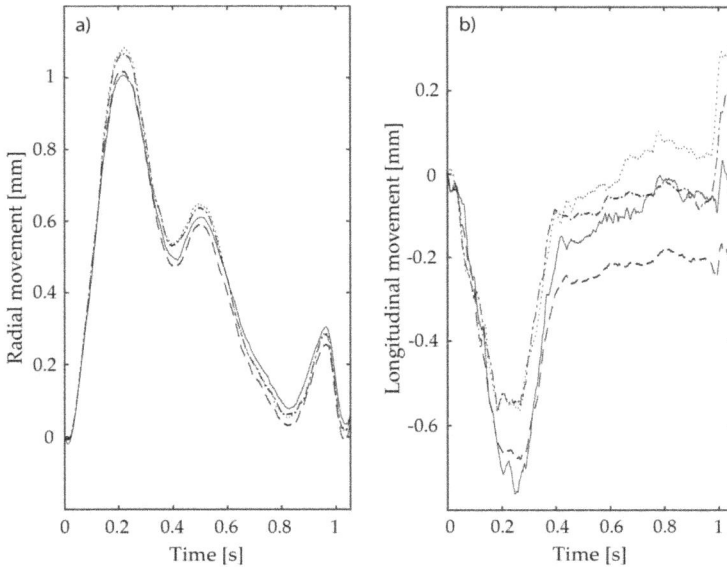

**Figure 9.** Radial movement (**a**) and longitudinal movement (**b**) of the intima–media complex of the common carotid artery wall of a healthy 35-year-old female. The kernels for the solid, dashed, dash-dot, and dotted lines were placed in order from left to right, respectively, in the first frame cine loop. The head was to the left of the image and the heart to the right. The frames were sampled at 1500 Hz and the proposed tracking scheme had an initial length of iteration of $k = 32$ frames.

## 4. Discussion

Frame-to-frame tracking of lateral tissue motion of the arterial wall using ultrafast ultrasound imaging is difficult as the movement per frame (<15 μm) is very small compared to the line distance (100–200 μm). The relative motion estimation error increases with decreasing movement per frame and the resulting movement curves become noisy and unreliable. To overcome this problem, we propose an iterative tracking scheme, with a Lagrangian viewpoint, based on temporarily down-sampling the frame-rate, anteroposterior tracking, and unbiased block-matching using two kernels per position estimate. The proposed motion estimation scheme performed well in the phantom study when estimating both velocity and displacement. The results showed increased tracking accuracy using longer initial length of iterations ($k \geq 64$). The tracking performance was better using the 100 μm pitch transducer than the 200 μm pitch transducer. The limited in vivo study showed that the proposed 2D tissue motion estimation tracking scheme can be used in vivo and is one step towards pulse-wave velocity estimations of the longitudinal movement of the arterial wall.

As the relative motion estimation error increases with decreasing movement per frame [26], we hypothesized that it would be easier to accurately estimate a large motion than a small motion. To achieve a larger motion per frame, we used temporary down-sampling of the frame-rate. The temporary down-sampling of the frame rate also made it possible to perform anteroposterior tracking—tracking both forward and backward in the cine loop—and use two kernels per position estimate in the next iteration. The benefit of anteroposterior tracking is that bias of the motion estimate is reduced, and if the forward and the backward movements are equal, the bias is cancelled. We have previously shown that the use of an extra kernel per position reduces the mean tracking estimation errors [28]. The tracking scheme also uses much fewer estimations from the start of the tracking before reaching the investigated frame compared to frame-to-frame tracking. The combined effect of these factors is clearly shown in Figures 3, 6 and 7 as the accumulated motion estimation errors decrease with a larger initial length of iteration ($k$).

In this study, both the velocity and the displacement were evaluated. The rationale is that these parameters investigate different features of a tracking method: the velocity evaluation gives an indication of the instantaneous uncertainty, whereas the displacement gives an indication of the accumulated uncertainty. In this study, the standard deviations of the velocity estimates are seemingly large. However, it should be noted that a velocity error of 1.3 mm/s corresponds to a displacement error of 1 μm/frame (frame rate = 1302 Hz) and this is one reason why frame-to-frame tracking is likely to fail. In the evaluation, the velocity was estimated between two consecutive frames and only one estimate per kernel was used. Despite this, our proposed iterative tracking scheme based on temporarily down-sampling of the frame-rate and anteroposterior tracking enables the displacement to be accurately tracked (Figure 3, Tables 3 and 4).

Averaging of two independent sources is a common method to obtain a more robust measurement. Time-averaging over multiple frames can be performed but acts as a low-pass filter in the time domain and can hide vital information. The averaging in our method is fundamentally different in that it averages the positions of the kernels and not its movement. We achieve a more robust estimate of the position of the kernel in each frame without affecting the time resolution using an extra kernel per position estimate [28].

The signal-to-noise ratio, speckle decorrelation, out-of-plane movement, and biasing are important factors for the size of the estimation uncertainty. In ultrafast ultrasound imaging, the signal-to-noise ratio is the most important factor whereas the others mainly have an effect at larger movements per frame. We expect that the proposed tracking scheme will be affected by these factors and the size of the kernel in the same manner as other block-matching methods [26,28,31]. It is well known that larger kernel sizes give more robust motion estimations [31]. In the phantom study, we use a kernel size of $1 \times 1$ mm$^2$, which can be regarded to be a relatively small kernel size and might explain some of the standard deviations of the motion estimations in this study. These issues need further studies.

The main finding of the phantom study is that a large length of iteration (large $k$) reduces the mean estimation error and the standard deviation. However, the results are complex. Figures 4–7 show the following:

- Using a small initial length of iteration ($k \leq 2$) gave rather small mean estimation errors but gave large standard deviations. Each of the motion estimations in the first iteration gave a very small error, but they accumulated to rather large errors and did so along different paths.
- Using a medium initial length of iteration ($k = 4$–32) gave larger mean estimation errors but smaller standard deviations. All estimations were roughly equal, but the initial motion estimations underestimated the motions. The later iterations gave accurate estimations for the in-between frames, but their starting points from the first iteration were incorrect.
- Using a large initial length of iteration ($k \geq 64$) gave small mean estimation errors and small standard deviations. The distance moved between each frame in the first iteration was large enough for the motion estimations to be accurate and for the later iterations to give accurate estimations for the in-between frames.

It could be expected that the motion estimation errors were the same when using the initial length of iteration $k$ for a velocity $v$ or when using $k$ divided by 2 for a velocity two times $v$ as the movement per frame in the initial tracking is equal. One possible explanation for the different errors, shown in Figures 6 and 7, is the fact that an estimation using $k$ divided by two uses one iteration less than an estimation using $k$. Overall, as stated above, more iterations give less estimation errors in the phantom measurements; however, this issue needs further studies.

The motion estimations of the proposed tracking scheme were compared to state-of-the-art low frame rate motion estimates as this is the gold standard when estimating the longitudinal wall movement of the arterial wall [26,28,32–36]. The optimal frame rate using low frame rate imaging with a conventional linear scan scheme depends on, e.g., the signal-to-noise ratio, speckle decorrelation, out-of-plane movement, and biasing. Somewhat depending on the ultrasound scanner, we regard a frame rate of 50–90 Hz to be the optimal using a conventional linear scan scheme. Considering the large estimation errors when tracking frame-to-frame using ultrafast ultrasound imaging ($k = 1$ in Tables 1–4), the motion estimation errors when using initial length of iteration $k = 128$ are promising as they are of the same order of magnitude as the results using low frame rate cine loops.

The effect of using transducers with different pitch (Figures 4 and 6) was anticipated in effect if not in amplitude, i.e., tracking using a smaller pitch ($\lambda/2$) gives more accurate motion estimations than using the larger pitch ($\lambda$). This depends probably on both higher pixel density in the lateral direction in the resulting B-mode image and improved beamforming because of the smaller pitch. Further studies are needed to evaluate this.

The tracking performance using a low frame rate and the larger pitch was unexpectedly poor, raising the question of whether low frame rate tracking using the 200-μm-pitch transducer can work in vivo. However, the ultrasound scanner that was used on the phantom set-up is a research scanner without, e.g., virtual scan lines, whereas the scanner we used to perform state-of-the-art tracking in vivo is a commercial ultrasound scanner utilizing virtual scan lines and many other techniques that improve the image quality. Several studies have shown that the tracking performance is sufficient when using low frame rate imaging with a conventional linear scan scheme [26,28,32–36]. The frame rate in these studies varies between 30–100 Hz.

The benefit of the implemented iterative method came not only from the use of two kernels for each estimation but also from the length of the movement between the frames in the first iteration. Considering that the initial length of iteration should be "long enough" for the best tracking accuracy and that the velocities vary drastically in in vivo measurements, it is likely that the tracking performance of the proposed tracking scheme can be optimized in vivo by using an initial length of iteration that adapts to the tissue velocity aiming to achieve the longest possible movement without significant speckle decorrelation. Potentially the best results could be achieved by using two estimations of the movement in a cine loop: the first one determining the overall shape of the movement using a fixed frame rate, e.g., 50 Hz corresponding to $k = 32$, in order to maximize the initial length of iteration for each part of the cine loop; the second estimation determining the movement in all frames.

Tracking of the longitudinal movement of the arterial wall frame-to-frame using the combination block-matching and ultrafast ultrasound imaging does not work. Though no ground-truth exists, the major movement patterns in vivo using our proposed iterative tracking scheme correspond well with low frame rate motion estimates, indicating that the proposed tracking scheme could enable the study of longitudinal movement imaging of the intima–media complex using ultrafast ultrasound imaging. However, caution should be taken when drawing conclusions from these results as the magnitudes of the estimation errors on the in vivo measurements are yet unknown. In the subjects investigated, we obtained the best result using two different initial lengths of iteration ($k = 32$ and 64). Further studies are needed to individually optimize the $k$.

The tracking scheme presented here is easy to implement and we believe that it can be used with most motion estimation methods with a Lagrangian viewpoint. In our implementation, the computational load is the same for any $k$.

There are two limitations in the phantom study. The first limitation is the chosen constant velocities of the phantom. With a maximal velocity of 2 mm/s, we are well below fast tissue motions in vivo. The second limitation is that the largest initial length of iterations was $k = 128$. Larger values of $k$ were not possible to test due to the combination of using a Lagrangian viewpoint, the velocity of the phantom, and the size of the ultrasound frames. However, we do not know whether a continued increase of $k$ will be beneficial as speckle decorrelation and out-of-plane movement can be increasingly problematic. Also, the results from the in vivo measurement (Figures 8 and 9), where $k = 32$ and 64, were optimal, indicating that the proposed tracking scheme does not necessarily continue to improve with increased $k$ when used on in vivo measurements. Further studies, preferably with simulated data or a larger set of in vivo cine loops, are needed. Undesirable fluctuations, which are considered to be caused by pitching and yawing of the automatic stages and electrical noise in the measurement system, were contained in the estimated lateral and axial velocities. However, displacements due to such cyclic components, which are presumably caused by pitching and yawing of the automatic stages, were very small.

## 5. Conclusions

Ultrafast ultrasound imaging provides excellent time resolution of motion and enables visualization of fast processes such as the pulse wave propagation of the arterial wall. The radial pulse wave propagation has been visualized using ultrafast ultrasound imaging [14,15], but it has been more challenging to visualize the longitudinal movement and hence the propagation of the longitudinal movement of the arterial wall. A robust method for estimating 2D motions in ultrafast ultrasound cine loops is needed for estimation of the longitudinal movement, and here we have presented a tracking scheme that might fill that role. The phantom evaluation clearly shows that our tracking scheme reduced the accumulated errors. In addition, the limited in vivo study shows that the major movement patterns in vivo correspond well with low frame rate motion estimates, indicating that the proposed tracking scheme could enable the study of longitudinal movement of the intima–media complex using ultrafast ultrasound imaging, and is one step towards estimating the propagation velocity of the longitudinal movement of the arterial wall.

**Author Contributions:** J.A., H.T. and M.C. were involved in the design of the new method, J.A., H.H. and M.C. conceived and designed the experiments; H.H., J.A., Å.R.A., M.C. performed the experiments; J.A. analyzed the data; J.A., Å.R.A. and M.C. interpreted the results, H.H., E.B. and A.R. contributed with instrumentation and instrumentation knowledge; J.A. wrote the paper; All authors approved the final draft of the manuscript.

**Acknowledgments:** We thank Ann-Kristin Jönsson for skillful technical assistance and The Swedish Research Council, the Swedish Foundation for International Cooperation in Research and Higher Education (STINT), the Medical and Technical Faculties, Lund University, and the Skåne County Council's Research and Development Foundation for funding.

**Conflicts of Interest:** The authors declare no conflict of interest. The founding sponsors had no role in the design of the study; in the collection, analyses, or interpretation of data; in the writing of the manuscript, and in the decision to publish the results.

## References

1. Gennisson, J.-L.; Provost, J.; Deffieux, T.; Papadacci, C.; Imbault, M.; Pernot, M.; Tanter, M. 4-D Ultrafast Shear-Wave Imaging. *IEEE Trans. Ultrason. Ferroelectr. Freq. Control* **2015**, *62*, 1059–1065. [CrossRef] [PubMed]
2. Bercoff, J.; Chaffai, S.; Tanter, M.; Sandrin, L.; Catheline, S.; Fink, M.; Gennisson, J.L.; Meunier, M. In Vivo Breast Tumor Detection using Transient Elastography. *Ultrasound Med. Biol.* **2003**, *29*, 1387–1396. [CrossRef]
3. Muller, M.; Gennisson, J.-L.; Deffieux, T.; Tanter, M.; Fink, M. Quantitative Viscoelasticity Mapping of Human Liver using Supersonic Shear Imaging: Preliminary in Vivo Feasibility Study. *Ultrasound Med. Biol.* **2009**, *35*, 219–229. [CrossRef] [PubMed]
4. Tanter, M.; Bercoff, J.; Sandrin, L.; Fink, M. Ultrafast Compound Imaging for 2-D Motion Vector Estimation: Application to Transient Elastography. *IEEE Trans. Ultrason. Ferroelectr. Freq. Control* **2002**, *49*, 1363–1374. [CrossRef] [PubMed]
5. Palmeri, M.L.; Wang, M.H.; Dahl, J.J.; Frinkley, K.D.; Nightingale, K. Quantifying Hepatic Shear Modulus in Vivo using Acoustic Radiation Force. *Ultrasound Med. Biol.* **2008**, *34*, 546–558. [CrossRef] [PubMed]
6. Hansen, P.M.; Olesen, J.B.; Pihl, M.J.; Lange, T.; Heerwagen, S.; Pedersen, M.M.; Rix, M.; Lönn, L.; Jensen, J.A.; Nielsen, M.B. Volume Flow in Arteriovenous Fistulas using Vector Velocity Ultrasound. *Ultrasound Med. Biol.* **2014**, *40*, 2707–2714. [CrossRef] [PubMed]
7. Lenge, M.; Ramalli, A.; Boni, E.; Liebgott, H.; Cachard, C.; Tortoli, P. High-frame-rate 2-D vector blood flow imaging in the frequency domain. *IEEE Trans. Ultrason. Ferroelectr. Freq. Control* **2014**, *61*, 1504–1514. [CrossRef] [PubMed]
8. Takahashi, H.; Hasegawa, H.; Kanai, H. Echo speckle imaging of blood particles with high-frame-rate echocardiography. *Jpn. J. Appl. Phys.* **2014**, *53*, 07KF08. [CrossRef]
9. Deffieux, T.; Jean-Luc, G.; Tanter, M.; Fink, M. Assessment of the Mechanical Properties of the Musculoskeletal System Using 2-D and 3-D Very High Frame Rate Ultrasound. *IEEE Trans. Ultrason. Ferroelectr. Freq. Control* **2008**, *55*, 2177–2190. [CrossRef] [PubMed]
10. Errico, C.; Osmanski, B.-F.; Pezet, S.; Couture, O.; Lenkei, Z.; Tanter, M. Transcranial functional ultrasound imaging of the brain using microbubble-enhanced ultrasensitive Doppler. *NeuroImage* **2016**, *124*, 752–761. [CrossRef] [PubMed]
11. Tong, L.; Gao, H.; Choi, H.F.; D'hooge, J. Comparison of Conventional Parallel Beamforming With Plane Wave and Diverging Wave Imaging for Cardiac Applications: A Simulation Study. *IEEE Trans. Ultrason. Ferroelectr. Freq. Control* **2012**, *59*, 1654–1663. [CrossRef] [PubMed]
12. Hasegawa, H.; Kanai, H. High-frame-rate echocardiography using diverging transmit beams and parallel receive beamforming. *J. Med. Ultrason.* **2011**, *38*, 129–140. [CrossRef] [PubMed]
13. Hasegawa, H.; Kanai, H. Simultaneous imaging of artery-wall strain and blood flow by high frame rate acquisition of RF signals. *IEEE Trans. Ultrason. Ferroelectr. Freq. Control* **2008**, *55*, 2626–2639. [CrossRef] [PubMed]
14. Salles, S.; Chee, A.J.Y.; Garcia, D.; Yu, A.C.H.; Vray, D.; Liebgott, H. 2-D Arterial Wall Motion Imaging Using Ultrafast Ultrasound and Transverse Oscillations. *IEEE Trans. Ultrason. Ferroelectr. Freq. Control* **2015**, *62*, 1047–1058. [CrossRef] [PubMed]
15. Kruizinga, P.; Mastik, F.; van den Oord, S.C.H.; Schinkel, A.F.L.; Bosch, J.G.; de Jong, N.; van Soest, G.; van der Steen, A.F.W. High-Definition Imaging of Carotid Artery Wall Dynamics. *Ultrasound Med. Biol.* **2014**, *40*, 2392–2403. [CrossRef] [PubMed]
16. Nichols, W.W.; O'Rourke, M.F. *McDonald's Blood Flow in Arteries*, 6th ed.; Edward Arnold: London, UK, 2011.
17. Blacher, J.; Guerin, A.P.; Pannier, B.; Marchais, S.J.; Safar, M.E.; London, G.M. Impact of aortic stiffness on survival in endstage renal disease. *Circulation* **1999**, *99*, 2434–2439. [CrossRef] [PubMed]
18. Cinthio, M.; Ahlgren, Å.R.; Bergkvist, J.; Jansson, T.; Persson, H.W.; Lindström, K. Longitudinal movements and resulting shear strain of the arterial wall. *Am. J. Physiol. Heart. Circ. Physiol.* **2006**, *291*, H394–H402. [CrossRef] [PubMed]
19. Nilsson, T.; Ahlgren, Å.R.; Jansson, T.; Persson, H.W.; Nilsson, J.; Lindström, K.; Cinthio, M. A method to measure shear strain with high-spatial-resolution in the arterial wall non-invasively in vivo by tracking zerocrossings of B-Mode intensity gradients. In Proceedings of the 2010 IEEE Ultrasonics Symposium (IUS), San Diego, CA, USA, 11–14 October 2010; pp. 491–494.

20. Idzenga, T.; Holewijn, S.; Hansen, H.H.G.; de Korte, C.L. Estimating Cyclic Shear Strain in the Common Carotid Artery Using Radiofrequency Ultrasound. *Ultrasound Med. Biol.* **2012**, *38*, 2229–2237. [CrossRef] [PubMed]

21. Zahnd, G.; Boussel, L.; Serusclat, A.; Vray, D. Intramural shear strain can highlight the presence of atherosclerosis: A clinical in vivo study. In Proceedings of the 2011 IEEE International Ultrasonics Symposium (IUS), Orlando, FL, USA, 18–21 October 2011; pp. 1770–1773.

22. Ahlgren, Å.R.; Cinthio, M.; Steen, S.; Nilsson, T.; Sjöberg, T.; Persson, H.W.; Lindström, K. Longitudinal displacement and intramural shear strain of the porcine carotid artery undergo profound changes in response to catecholamines. *Am. J. Physiol. Heart. Circ. Physiol.* **2012**, *302*, H1102–H1115. [CrossRef] [PubMed]

23. Zahnd, G.; Maple-Brown, L.J.; O'Dea, K.; Moulin, P.; Celermajer, D.S.; Skilton, M.R.; Vray, D.; Sérusclat, A.; Alibay, D.; Bartold, M.; et al. Longitudinal displacement of the carotid wall and cardiovascular risk factors: Associations with aging, adiposity, blood pressure and periodontal disease independent of cross-sectional distensibility and intima-media thickness. *Ultrasound Med. Biol.* **2012**, *38*, 1705. [CrossRef] [PubMed]

24. Svedlund, S.; Eklund, C.; Robertsson, P.; Lomsky, M.; Gan, L.-M. Carotid artery longitudinal displacement predicts 1-year cardiovascular outcome in patients with suspected coronary artery disease. *Arterioscler. Thromb. Vasc. Biol.* **2011**, *31*, 1668–1674. [CrossRef] [PubMed]

25. Svedlund, S.; Gan, L.-M. Longitudinal common carotid artery wall motion is associated with plaque burden in man and mouse. *Atherosclerosis* **2011**, *217*, 120–124. [CrossRef] [PubMed]

26. Albinsson, J.; Ahlgren, Å.R.; Jansson, T.; Cinthio, M. A combination of parabolic and grid slope interpolation for 2D tissue displacement estimations. *Med. Biol. Eng. Comput.* **2017**, *55*, 1327–1338. [CrossRef] [PubMed]

27. Hasegawa, H. Phase-Sensitive 2D Motion Estimators Using Frequency Spectra of Ultrasonic Echoes. *Appl. Sci.* **2016**, *6*, 195. [CrossRef]

28. Albinsson, J.; Brorsson, S.; Ahlgren, Å.R.; Cinthio, M. Improved Tracking Performance of Lagrangian Block-Matching Methodologies using Block Expansion in the Time Domain—In silico, phantom and in vivo evaluations using ultrasound images. *Ultrasound Med. Biol.* **2014**, *40*, 2508–2520. [CrossRef] [PubMed]

29. Boni, E.; Bassi, L.; Dallai, A.; Guidi, F.; Ramalli, A.; Ricci, S.; Housden, J.; Tortoli, P. A reconfigurable and programmable FPGA-based system for nonstandard ultrasound methods. *IEEE Trans. Ultrason. Ferroelectr. Freq. Control* **2012**, *59*, 1378–1385. [CrossRef] [PubMed]

30. Tortoli, P.; Bassi, L.; Boni, E.; Dallai, A.; Guidi, F.; Ricci, S. An Advanced Open Platform for ULtrasound Research. *IEEE Trans. Ultrason. Ferroelectr. Freq. Control* **2009**, *56*, 2207–2216. [CrossRef] [PubMed]

31. Friemel, B.H.; Bohs, L.N.; Trahey, G.E. Relative performance of two-dimensional speckle-tracking techniques: Normalized correlation, non-normalized correlation and sum-absolute-difference. *Proc. IEEE Ultrason.* **1995**, *2*, 1481–1484.

32. Cinthio, M.; Ahlgren, Å.R.; Jansson, T.; Eriksson, A.; Persson, H.W.; Lindström, K. Evaluation of an ultrasonic echo-tracking method for measurements of arterial wall movements in two dimensions. *IEEE Trans. Ultrason. Ferroelectr. Freq. Control* **2005**, *52*, 1300–1311. [CrossRef] [PubMed]

33. Cinthio, M.; Ahlgren, Å.R. Intra-Observer Variability of Longitudinal Movement and Intramural Shear Strain Measurements of the Arterial Wall using Ultrasound Non-Invasively in vivo. *Ultrasound Med. Biol.* **2010**, *36*, 697–704. [CrossRef] [PubMed]

34. Zahnd, G.; Boussel, L.; Marion, A.; Durand, M.; Moulin, P.; Serusclat, A.; Vray, D. Measurement of Two-Dimensional Movement Parameters of the Carotid Artery Wall for Early Detection of Arteriosclerosis: A Preliminary Clinical Study. *Ultrasound Med. Biol.* **2011**, *37*, 1421–1429. [CrossRef] [PubMed]

35. Numata, T.; Hasegawa, H.; Kanai, H. Basic study on detection of outer boundary of arterial wall using its longitudinal motion. *Jpn. J. Appl. Phys.* **2007**, *46*, 4900–4907. [CrossRef]

36. Yli-Ollila, H.; Laitinen, T.; Weckström, M.; Laitinen, T.M. Axial and radial waveforms in Common Carotid Artery: An advanced method for studying arterial elastic properties in ultrasound imaging. *Ultrasound Med. Biol.* **2013**, *39*, 1168–1177. [CrossRef] [PubMed]

![applied sciences logo] *applied sciences*

MDPI

Article

# Effect of Ultrafast Imaging on Shear Wave Visualization and Characterization: An Experimental and Computational Study in a Pediatric Ventricular Model

**Annette Caenen [1],\*, Mathieu Pernot [2], Ingvild Kinn Ekroll [3], Darya Shcherbakova [1], Luc Mertens [4], Abigail Swillens [1] and Patrick Segers [1]**

[1] IBiTech-bioMMeda, Ghent University, 9000 Ghent, Belgium; darya.shcherbakova@ugent.be (D.S.); abbyswillens@gmail.com (A.S.); patrick.segers@ugent.be (P.S.)
[2] Institut Langevin, Ecole Supérieure de Physique et de Chimie Industrielles, CNRS UMR 7587, INSERM U979, 75012 Paris, France; mathieu.pernot@gmail.com
[3] Circulation and Medical Imaging, Norwegian University of Science and Technology, 7491 Trondheim, Norway; ingvild.k.ekroll@ntnu.no
[4] Hospital for Sick Children, University of Toronto, Toronto, ON M5G 1X8, Canada; luc.mertens@sickkids.ca
\* Correspondence: annette.caenen@gmail.com; Tel.: +32-93-324320

Received: 14 July 2017; Accepted: 12 August 2017; Published: 16 August 2017

**Featured Application: Clinical application of Shear Wave Elastography for cardiac stiffness assessment in children.**

**Abstract:** Plane wave imaging in Shear Wave Elastography (SWE) captures shear wave propagation in real-time at ultrafast frame rates. To assess the capability of this technique in accurately visualizing the underlying shear wave mechanics, this work presents a multiphysics modeling approach providing access to the true biomechanical wave propagation behind the virtual image. This methodology was applied to a pediatric ventricular model, a setting shown to induce complex shear wave propagation due to geometry. Phantom experiments are conducted in support of the simulations. The model revealed that plane wave imaging altered the visualization of the shear wave pattern in the time (broadened front and negatively biased velocity estimates) and frequency domain (shifted and/or decreased signal frequency content). Furthermore, coherent plane wave compounding (effective frame rate of 2.3 kHz) altered the visual appearance of shear wave dispersion in both the experiment and model. This mainly affected stiffness characterization based on group speed, whereas phase velocity analysis provided a more accurate and robust stiffness estimate independent of the use of the compounding technique. This paper thus presents a versatile and flexible simulation environment to identify potential pitfalls in accurately capturing shear wave propagation in dispersive settings.

**Keywords:** ultrafast imaging; shear wave elastography; multiphysics modeling

---

## 1. Introduction

Ultrafast ultrasound imaging uses plane-wave transmissions instead of the conventional line-by-line focused beam transmissions, increasing the frame rate by at least a factor of 100 (typically >1000 frames per second) [1,2]. This ultrafast imaging technology was an essential breakthrough for the field of Shear Wave Elastography (SWE), as it allowed real-time imaging of shear waves in soft tissues with a high temporal resolution [3–5]. Because of this, the technique was almost instantaneously applied and therefore less sensitive to respiratory and/or cardiac motion. This allowed local quantitative estimates of wave speed and therefore of tissue stiffness [6]. Initially, shear waves were generated with a transient vibration originating from an external mechanical vibrator [3,4]. However, as these vibrators were challenging to integrate in daily clinical practice, the excitation source was changed into a remote palpation induced by a radiation force of focused ultrasonic beam(s), unifying the shear wave excitation source and ultrafast imaging modality together in the ultrasound transducer [5,7–9]. At the beginning, the ultrafast frame rates came at the cost of reduced image contrast and resolution compared to conventional transmissions as the transmit focusing step is skipped in the ultrafast imaging modality. However, this limitation was overcome by introducing coherent plane wave compounding [4,10], which consists of sending out multiple tilted and non-tilted plane waves into the medium and coherently summing the backscattered echoes to compute the full image. In this manner, the image quality is improved compared to single plane wave imaging while still maintaining sufficiently high frame rates [10]. The concept of compounding has been applied to different ultrasound modalities [11–14], and has become a key feature of ultrafast ultrasound imaging.

Ultrafast imaging in SWE to assess tissue stiffness has been clinically applied in several areas such as breast cancer diagnosis [15] and liver fibrosis staging [16]. The ability of ultrafast imaging—with or without plane wave compounding—in displaying and characterizing the true biomechanical shear wave propagation has not been well studied yet, to the best of our knowledge. We are particularly interested in the performance of ultrafast imaging in tissues with thin and layered geometries and other intricate anisotropic material properties, as complex shear wave propagation phenomena such as wave guiding, mode conversions and dispersion are expected to arise [17,18]. These wave features will complicate shear wave visualization, characterization and interpretation, eventually affecting SWE-based stiffness estimation. This may be especially true when plane wave compounding is applied, as the compounded image fuses temporal characteristics of the propagating shear wave at different time points. Indeed, a recent study in ex vivo thoracic aorta [19] has experimentally shown that certain SWE settings, such as pushing length and number of compounding angles, influenced the technique's accuracy to estimate phase velocity-based tissue stiffness.

Therefore, the objective of this work was to establish a flexible framework that allows us to investigate the performance of ultrafast imaging in SWE in accurately displaying and characterizing the true biomechanical shear wave propagation. As actual SWE experiments do not provide access to a ground truth for imaged shear wave propagation, a multiphysics modeling approach combining computational solid mechanics (CSM) of the shear wave propagation [20–22] with ultrasound (US) modeling of ultrafast imaging was used for this purpose. The resulting wave mechanics from CSM provided the true mechanical shear wave propagation whereas the virtual images represented the imaged shear wave propagation. The multiphysics model was employed in combination with SWE experiments, for validation purposes. This combined approach was applied on an idealized left ventricular phantom model with pediatric dimensions, as this has been demonstrated to evoke dispersive guided wave propagation patterns due to left ventricular geometry [23]. The proposed multiphysics model in this work thus adds an extra modeling layer to the previously presented SWE biomechanics model in [23], expanding our scope from studying the effect of *biomechanical* factors on shear wave physics to investigating the effect of *imaging* factors on shear wave physics. Our objective can be translated into two main study questions: (i) study the effect of compounding through comparison of single and compounded plane wave acquisitions from SWE experiments, for which more in-depth insights are realized by modeling both acquisitions using the multiphysics methodology, and (ii) study the effect of ultrafast imaging by analyzing the mechanical versus imaged shear wave acquisitions in the simulations. The study of each effect consisted of examining the shear wave propagation patterns in the time and frequency domain, and inspecting the accuracy of two different shear modulus estimation techniques, based on group and phase velocity, through comparison with the mechanically determined shear modulus.

*Appl. Sci.* **2017**, *7*, 840

## 2. Materials and Methods

### 2.1. SWE Experiments

SWE acquisitions were performed on an ultrasound phantom (10% polyvinylalcohol (PVA), freeze-thawed once) of the mimicking pediatric left ventricular geometry as illustrated in Figure 1. Further details on this phantom can be found in a recent publication from our group [23]. Shear waves were generated and imaged by a SL15-4 linear transducer with 256 elements, a pitch of 200 μm and an elevation focus of ~30 mm, connected to the Aixplorer system (SuperSonic Imagine, Aix-en-Provence, France). We considered two SWE acquisitions, one with single plane wave emissions (0°) and the second with coherent plane wave compounding ($-2°$, $0°$, $2°$) [10], in which the single plane waves are emitted at a pulse repetition frequency (PRF) of 6.9 kHz for both acquisitions. All other pushing and imaging parameters for both SWE acquisitions are listed in Table 1. The Aixplorer system provided us beamformed in-phase and quadrature-demodulated (IQ) signals with a fast time sampling rate of 32 MHz.

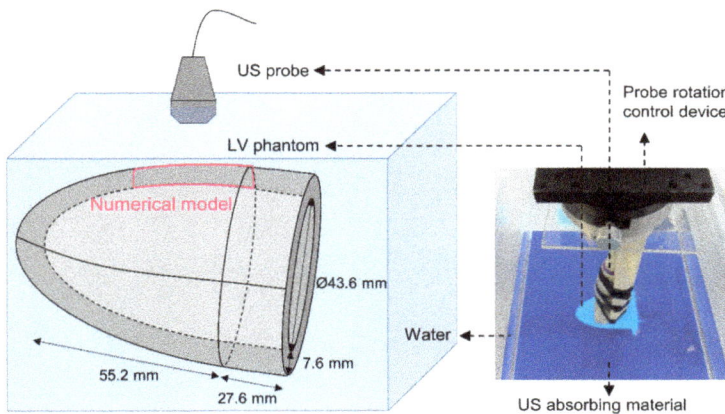

**Figure 1.** Experimental set-up (dimensions are not to scale in schematic diagram); US: ultrasound; LV: left ventricle.

**Table 1.** In vitro imaging parameters.

|  | Parameters | Values |
|---|---|---|
| Pushing sequence | Push frequency $f_0$ | 8 MHz |
|  | F-number | 2.5 |
|  | Apodization | - |
|  | Push duration | 250 μs |
| Imaging sequence | Number of cycles | 2 |
|  | Emission frequency | 8 MHz |
|  | Pulse repetition frequency (PRF) | 6.9 kHz |
|  | Imaging depth | 40 mm |
|  | F-number on transmit | - |
|  | Transmit apodization | - |
|  | F-number on receive | 1.2 |
|  | Receive apodization | Hanning |
|  | Receive bandwidth | 60% |

### 2.2. SWE Multiphysics Model

Concordant with an actual SWE measurement, the SWE model also splits the SWE acquisition into a pushing and an imaging sequence. This multiphysics platform contains three modeling parts, i.e., modeling of the acoustic radiation force (ARF), the shear wave propagation and the ultrafast imaging acquisition. The first two modeling parts compose the pushing sequence, whereas the third modeling part represents the imaging sequence (see Figure 2). These models need to be run consecutively as the output of the first model is used as input for the second model and likewise for the second and third model, as indicated by the arrows in Figure 2. The first and the third part of the multiphysics platform model the ultrasound physics through Field II [24,25], whereas the second modeling part simulates the wave mechanics in the finite element software Abaqus (Abaqus Inc., Providence, RI, USA). The modeling methodology for the pushing and imaging sequence is concisely described below. The reader is referred to [23] for further details about the pushing sequence, comprising the first two modeling parts.

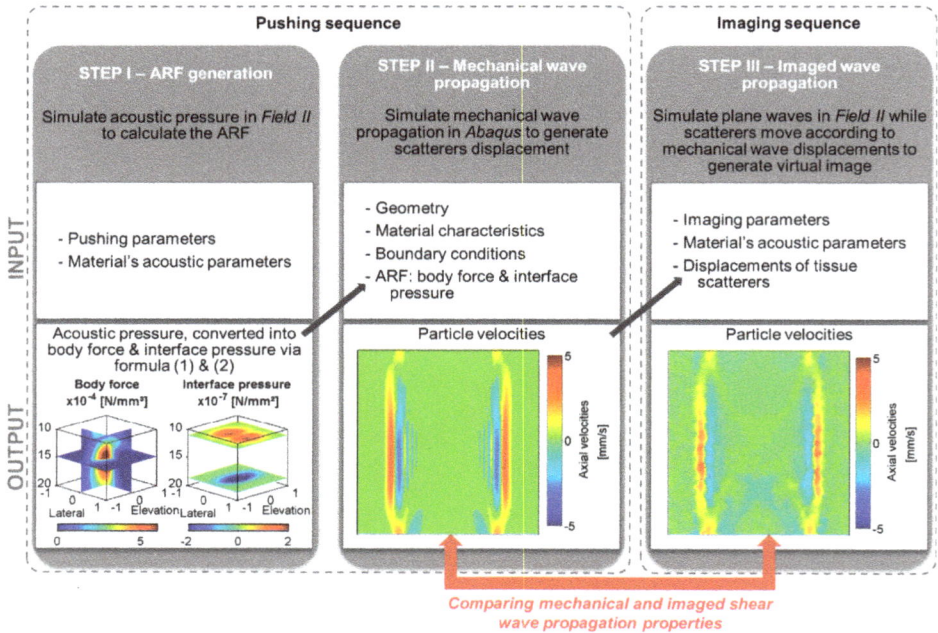

**Figure 2.** Workflow of the multiphysics platform; ARF: acoustic radiation force.

### 2.2.1. Pushing Sequence

The pushing sequence in the numerical model consists of two steps: ARF generation and mechanical wave propagation (Figure 2). For the first step, the ARF applied on the PVA phantom is numerically mimicked by a volume force in combination with an interface pressure. Both types of loading act on the PVA phantom in the focal zone of the probe, extending ~2 mm from the probe's center point in the lateral and elevation direction. The volume force acts throughout the complete thickness of the PVA phantom in this focal region, whereas the interface pressure is only active on the interfaces between phantom and water. Volume force $b$ and interface pressure $\pi$ are calculated based on the time-averaged acoustic intensity $I$, of which its spatial distribution is derived by simulating acoustic probe pressures mimicking the push sequence (see Table 1) with Field II and its magnitude is scaled to 1500 W/cm$^2$ [26], as follows [21,27]:

$$b = \frac{2\alpha I}{\rho c_L},$$ (1)

$$\pi = \frac{I}{c_1}\left(1 + R - (1-R)\frac{c_1}{c_2}\right),\qquad(2)$$

where $\alpha$ is the attenuation coefficient [dB/cm/MHz], $\rho$ the density of PVA [kg/m$^3$], $c_L$ the longitudinal wave speed of PVA [m/s], $R = \left(\frac{Z_2-Z_1}{Z_2+Z_1}\right)^2$ the energetic reflection coefficient [-], $Z_1$ and $Z_2$ the acoustic impedances ($Z_i = \rho_i c_i$) [Pa·s/m$^3$], and $c_1$ and $c_2$ the speeds of sound in media 1 and 2 [m/s]. Material characteristics of the modeled water and PVA can be found in Table 2. The PVA's Young's modulus and viscoelastic behavior were mechanically determined on a uniaxial tensile testing machine (Instron 5944, Norwood, MA, USA), whereas its density and speed of sound were measured using the principle of Archimedes [28] and an oscilloscope respectively (for more details on all measurements, we refer to [23]). The resulting spatial distribution of the volume force in the axial-lateral plane is shown in the bottom-left panel of Figure 2. Both loads are imposed for 250 µs in the numerical model.

Table 2. Material characteristics of water and polyvinylalcohol (PVA).

|  | Characteristics | Value |
|---|---|---|
| **Water** | Density $\rho$ | 1000 kg/m$^3$ |
|  | Speed of sound $c_L$ [29] | 1480 m/s |
|  | Bulk modulus $K$ [29] | 2200 MPa |
| **PVA** | Density $\rho$ | 1045.5 kg/m$^3$ |
|  | Speed of sound $c_L$ | 1568 m/s |
|  | Young's Modulus $E$ | 73.0 kPa |
|  | Attenuation coefficient $\alpha$ [30] | 0.4 dB/cm/MHz |
|  | Coefficient of Poisson $\nu$ [20] | 0.49999 |
|  | Normalized shear modulus $g_1$ | $4.04 \times 10^{-3}$ |
|  | Relaxation time $\tau_1$ | $99.8 \times 10^{-6}$ s |
|  | Normalized shear modulus $g_2$ | $7.04 \times 10^{-2}$ |
|  | Relaxation time $\tau_2$ | 77.9 s |

For the actual mechanical wave simulation (step II in Figure 2), the PVA phantom was modeled as one half of an ellipsoidal-shaped disk with a lateral and elevational length of 27.8 mm and 16.0 mm respectively, taking the symmetry of the imaging plane into account. For reasons of computational efficiency, we considered only half the width of the transducer in the model, and modeled structural infinite elements at the edges of the defined domain. This PVA model was meshed with 8-noded brick elements with reduced integration, leading to 355,680 elements in total. The water below and above the phantom is represented by two layers of 8-noded hexahedral acoustic elements, each with a thickness of 3.8 mm and 79,684 elements. Mechanical displacements of the PVA phantom were coupled to acoustic pressures in the water layer through a tie-constraint. The other surfaces of the modeled water were modeled to be infinite. The PVA was modeled as a viscoelastic material by assuming a 2-term Prony series model with normalized shear moduli $g_i$ and relaxation times $\tau_i$ as mentioned in Table 2, which are derived from a uniaxial mechanical relaxation test stretching the PVA material at 5% strain for 10 s [23]. It should be noted that the modeled viscoelasticity has a negligible influence on shear wave propagation characteristics, indicating that the actual and modeled PVA phantom have very low viscosity [23]. The water was defined as an acoustic medium in the model with bulk modulus $K$ and density $\rho$ as tabulated in Table 2. More details about mesh geometry, boundary conditions, material characteristics and loading can be found in [23].

The dynamic equations of motion of this numerical problem were solved by the Abaqus explicit solver and the particle velocities were extracted at a sampling rate of 40 kHz for further analysis. The wave propagation resulting from these simulations is called 'mechanical shear wave propagation' throughout this work (see Figure 2).

2.2.2. Imaging Sequence

The imaging sequence simulation in the multiphysics approach is illustrated in step 3 (Figure 2). The basis of the ultrasound simulation is Field II, in which tissue is represented by a collection of random point scatterers reflecting the ultrasonic waves emitted by the modeled probe. For each emitted beam, the scatterer's position is updated based on the CSM extracted displacement fields, utilizing first a temporal interpolation from the CSM timescale to US timescale and subsequently spatial interpolation from CSM mesh grid to US scatterer grid. In order to obtain a proper random distribution of point scatterers within our numerical phantom, we used an algorithm based on the open-source software Visualization ToolKit (VTK) [31]. This algorithm first generates randomly distributed scatterers in a box surrounding the phantom's geometry and then removes the abundant scatterers outside the actual geometry based on geometric criteria of the scatterers relative to the phantom's surface [32]. Approximately 10 scatterers per resolution cell (with its size calculated based on receive F-number, transmit frequency and pulse length) were considered to ensure a Gaussian-distributed RF signal [33].

To mimic our SWE experiments (see Section 2.1), two ultrafast imaging settings were simulated, one with and one without coherent plane wave compounding, using the same probe parameters as mentioned in Table 1. However, the virtual transducer's size was reduced to 128 piezoelectric elements to decrease computational time. For the same reason, the number of simulated frames was limited to 27 and 9 for the single and compounded Plane Wave Imaging (PWI) acquisition, respectively. For the estimation of the scatterer displacement during these simulations, the displacement information of the same CSM simulation was used since the pushing parameters or location did not change throughout the experiments (see Table 1). In our simulation setup, each transducer element was divided into four rectangular mathematical elements in the elevational direction to ensure a far-field approximation of the spatial impulse response. Channel data were acquired at a fast time sampling rate of 100 MHz, IQ-demodulated to 32 MHz and subsequently delay-and-sum beamformed with parameters mentioned in Table 1 using an in-house developed code from the Norwegian University of Science and Technology (NTNU). The obtained wave propagation from these simulations is termed 'virtual imaged shear wave propagation' throughout this work (see Figure 2).

*2.3. Post-Processing*

The data acquired from the SWE experiments and the SWE multiphysics model were both processed as described below to obtain the axial particle velocities as a function of time and space and the shear modulus estimate.

2.3.1. Axial Velocity Estimation

Axial velocities $\hat{v}_z$ were obtained by applying the autocorrelation technique on the IQ-data as follows [34,35]:

$$\hat{v}_z = \frac{c_L\left(\frac{PRF}{n_T}\right)}{4\pi f_0} \angle \hat{R}_x(1) \tag{3}$$

where $\angle \hat{R}_x(1)$ represents the phase angle of the autocorrelation function of lag one which is estimated from the received signal sequence, and $n_T$ the number of transmit beams to obtain one image. The axial velocity estimate was further improved by spatial averaging the autocorrelation estimate over an area of approximately $0.6 \times 0.6$ mm both in simulations and in vitro.

Note that this post-processing step is not applied on the mechanical shear wave simulations, as these immediately provide access to all components of the particle velocities in the 3D spatial domain. Additionally, the mechanical wave simulations have a slow time sampling rate of 40 kHz, whereas the sampling rate of the real and virtual SWE imaging measurements depends on the acquisition, i.e., 6.9 kHz for single plane wave emissions and 2.3 kHz for plane wave compounding.

2.3.2. Shear Modulus Estimation

As our previous work [23] has shown that dispersive shear wave propagation patterns arose in the studied setting due to geometry, two different shear modulus estimation techniques were applied on the mechanical and (real and virtual) imaged wave propagation, i.e., a time-of-flight (TOF) method—implemented in commercial SWE systems and used for non-dispersive media—and a phase velocity analysis—used for dispersive media. The real and virtual imaging acquisitions were pre-processed by averaging the axial velocities over 0.6 mm axial depth and temporally up-sampling the slow time domain by a factor 10.

For the TOF method, the shear wave's position was tracked by searching the maximal axial velocity for every lateral spatial location as a function of time and fitting a linear model to estimate the shear wave velocity (goodness of fit should be equal to or larger than 0.95) [23,36]. In general, to make the most complete use of the measured data and to increase the reliability of the fit, axial velocity data acquired from all probe elements should be taken into account during this linear fitting procedure. This is true for large isotropic homogeneous elastic media, but usually data from the probe's edge elements is discarded due to low signal-to-noise ratio and/or high attenuation of the propagating shear wave in the measurement. Even though the studied PVA setting is isotropic, homogeneous and low viscous, the left ventricular geometry induces dispersive shear wave features in the SWE-acquisitions which affect the tracked shear wave's position as a function of time. To investigate the effect of this observation on the results of the TOF method, we altered the number of data points taken into account during the TOF fitting procedure: the shear wave speed was estimated by rejecting 5 and 20 data points from the probe's edge elements for each shear wave. The shear modulus $\mu$ can then be derived from this wave speed $c_T$ by assuming an isotropic bulky elastic material with density $\rho$ and applying the following formula:

$$\mu = \rho c_T^2 \tag{4}$$

For the phase velocity analysis, measured or simulated dispersion characteristics were derived by taking the 2D Fast Fourier Transform (FFT) of the axial velocity wave propagation pattern as a function of lateral space and slow time at a specific depth [37]. Subsequently, the wavenumber $k$ with the maximal Fourier energy is tracked at each frequency $f$ in order to identify the main excited mode. Phase velocity $c_\varphi$ as a function of frequency $f$ is found through $c_\varphi = (2\pi f)/k$. The shear modulus is then estimated by fitting a theoretical model in a least squares manner to the obtained dispersion curve. Neglecting the ventricular curvature [38] and PVA's viscoelasticity, and assuming that the main excited mode is the first antisymmetric mode (A0) [17], we minimized the difference between the theoretical A0 dispersion curve of a plate in water and the extracted dispersion characteristics over a frequency range spanning from 0.2 kHz up to maximally 2 kHz, dependent on the considered acquisition [37,39]. Only fits giving a standard deviation less than 0.6 kPa for the shear modulus estimate were considered.

Both procedures were repeated for multiple depths across the phantom's thickness ($n = 10$). For further details on both shear modulus estimation techniques, we refer to [23].

## 3. Results

### 3.1. Analyzing the Shear Wave's Characteristics in the Time Domain

To study the shear wave's temporal characteristics, we examined its magnitude and shape throughout time by visualizing the axial velocities at three different time points. The resulting shear wave propagation of the experimentally measured SWE acquisitions with and without compounding are compared in Figure 3. Immediately, we observe a different shear wave propagation pattern: the shear wave front, represented by the downward axial velocities, is split into two for the single plane wave images whereas one uniform wave front is present for the compounded images. Furthermore, the wave front is also broader along the lateral direction when including compounding. Next to these differences in shear wave shape, we also observe a lower shear wave magnitude (maximal axial velocity amplitude at a certain time point can be up to 3 mm/s smaller) for the compounded images.

The imaged and biomechanical shear wave propagation for the simulations are depicted in Figure 4. For the biomechanical simulation (first row in Figure 4), we observe again a split in shear wave front during wave propagation, which is well captured in the virtual single plane wave images (second row in Figure 4), but less visible in the virtual compounded images (third row in Figure 4). Furthermore, the shear wave front is apparently broader in the imaging simulations compared to the biomechanical simulation. Additionally, the simulated axial velocity patterns of the virtual images show a clear decrease in tissue velocity magnitude (~23.0% for single PWI and ~69.4% for compounded PWI at the top of the phantom compared to the biomechanics simulation).

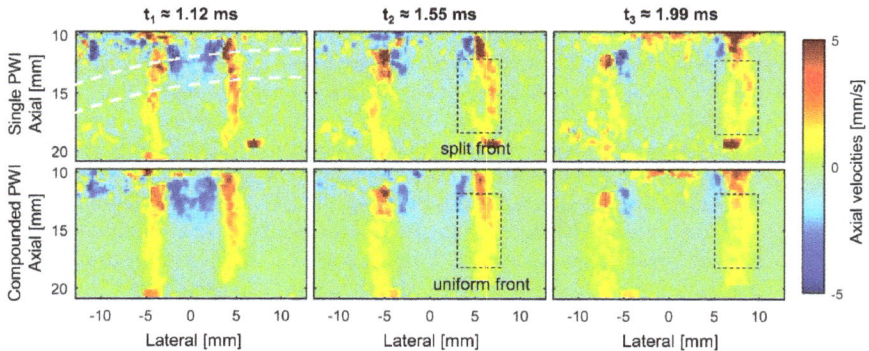

**Figure 3.** Comparison of shear wave propagation measured on the ventricular phantom at time points 1.12 ms, 1.55 ms and 1.99 ms (assuming $t_0 = 0$ s corresponds with the start of the pushing sequence and an ultrasound system's electronic dead time of 0 s) for single and compounded Plane Wave Imaging (PWI). The white dotted lines represent shear wave propagation paths at 15% and 40% tissue depth with respect to the ventricular thickness.

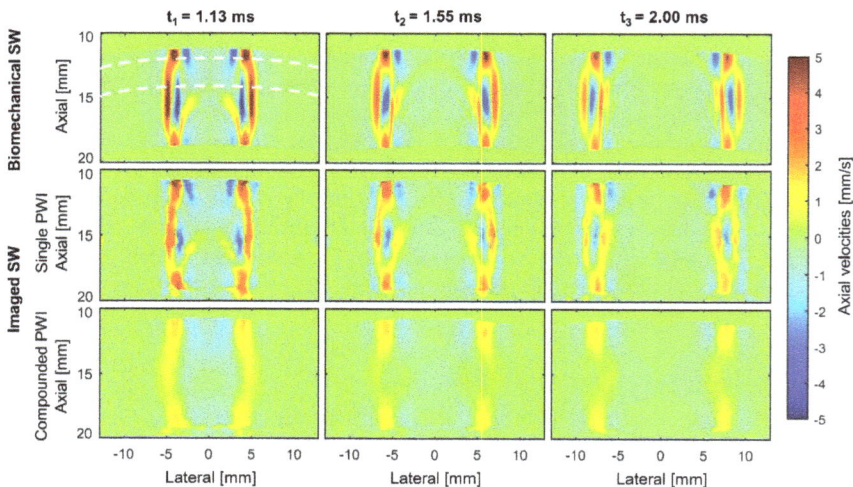

**Figure 4.** Comparison of the biomechanical shear wave propagation (upper panels) and the virtually imaged shear wave propagation without and with compounding (middle and lower panels respectively) at time points 1.13 ms, 1.55 ms and 2.00 ms (assuming $t_0 = 0$ s corresponds with the start of the pushing sequence). The white dotted lines represent shear wave propagation paths at 15% and 40% tissue depth with respect to the ventricular thickness.

## 3.2. Analyzing the Shear Wave's Characteristics in the Frequency Domain

The shear wave's frequency features were studied by taking the 2D FFT of the axial velocity map in time and lateral space (see Methods section) at 15% and 40% tissue thickness, representing two different shear wave propagation paths as indicated by the white dotted lines in Figures 3 and 4. The Fourier energy magnitudes of both simulations and measurements are mentioned in Table 3. Observations concerning mode(s) excitation, Fourier energy magnitude and frequency content in the Fourier spectra are consecutively discussed below.

**Table 3.** Tabulation of the magnitude of the maximal Fourier energy amplitude in Figure 5 [mm/s/Hz].

| Acquisition | | 15% Tissue Thickness | 40% Tissue Thickness |
|---|---|---|---|
| Experimental | | | |
| | Single PWI | 6.52 | 3.42 |
| | Compounded PWI | 1.68 | 1.26 |
| Numerical | | | |
| | Single PWI | 5.13 | 3.45 |
| | Compounded PWI | 0.64 | 0.33 |
| | Biomechanics | 33.37 | 31.91 |

### 3.2.1. Mode(s) Excitation

For experimental single PWI (first column of Figure 5), we observed that mainly one mode was excited at the shallow tissue depth, whereas two modes were excited for deeper tissue regions. The mode excited on lower frequencies is designated with the term 'primary mode', whereas the other mode is defined as 'secondary mode'. This primary mode is the one that will be tracked and fitted to the theoretical A0-mode in the phase velocity analysis to estimate shear stiffness. Applying compounding in the experiment led to one visible excited mode in the spectra of both tissue depths, as can be seen in the second column of Figure 5. For the simulations (third, fourth and fifth columns of Figure 5), we see one excited mode for 15% tissue thickness, and two excited modes for 40% tissue thickness, independent of the application of the compounding technique.

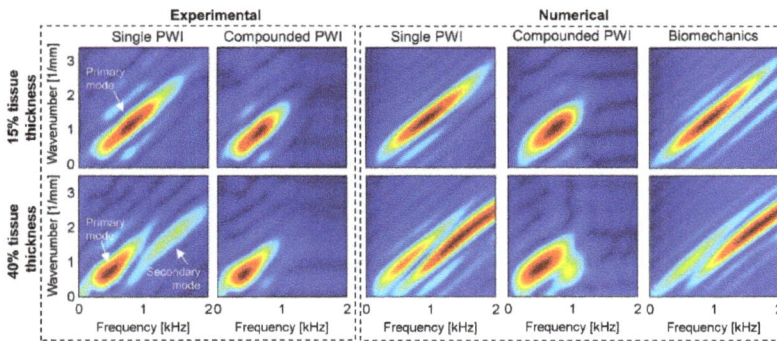

**Figure 5.** Fourier energy maps at two paths across the phantom's thickness—15% and 40%—for the right shear wave in the experimental (single and compounded PWI) and numerical (single PWI, compounded PWI and biomechanics) shear wave acquisitions. Location of the two shear wave paths is indicated in Figures 3 and 4 for experiment and simulation respectively. The primary mode is defined as the mode excited on lower frequencies and the secondary mode is the mode excited on higher frequencies, as indicated in the biomechanics column. Each Fourier energy map was normalized to its maximal energy (displayed in red); amplitudes are given in Table 3. The measured temporal shear wave data for one specific shear wave path across axial depth were cropped in lateral space (12.8 mm) and time (4 ms) such that its spatial and temporal resolution corresponded to the simulated ones.

3.2.2. Fourier Energy Magnitude

For the applied experiments, coherent compounding decreased the maximal Fourier energy magnitude with a factor of 3.9 for 15% tissue depth and 2.7 for 40% tissue thickness (see Table 3). A similar observation was made for the simulations: compounding reduced the maximal Fourier energy magnitude by factors of 8.0 and 10.5 for 15% and 40% tissue thickness respectively. Furthermore, when comparing the virtual single PWI to the biomechanics simulation, an additional decrease by factors of 6.5 and 9.2 was noticed for the two considered tissue depths. Next to these dissimilarities in maximal Fourier energy magnitude, the relative energy magnitude of secondary to primary mode for the deeper tissue region also differed (see Figure 5). This proportion was 0.5 for the experimental single PWI. For the simulations, this ratio shifted from 1.7 for the virtual biomechanics to 1.3 for single PWI and 0.6 for compounded PWI.

3.2.3. Frequency Content

The bandwidth of the Fourier spectra was about 2.0 kHz for the experimental single PWI and 1.0 kHz for the compounded acquisition (Figure 5), at both tissue depths. On the other hand, the bandwidth of the simulated Fourier spectra of the single PWI acquisition was around 2.0 kHz for 15% tissue thickness, and 3.0 kHz for 40% tissue thickness. When compounding was applied in the simulations, the maximal excited frequency was reduced to nearly 1.0 kHz for both tissue depths. However, the bandwidth of the biomechanical Fourier spectra of both virtual imaging acquisitions was about 2.0 kHz and 3.5 kHz for 15% and 40% tissue thickness respectively.

The frequency content of the detected signal was further changed when compounding was used: the frequency with maximal Fourier energy content shifted from 0.69 kHz to 0.79 kHz for 15% tissue thickness and from 0.42 kHz to 0.47 kHz for 40% tissue thickness. For the virtual single PWI, the maximal Fourier energy was reached at 0.98 kHz for 15% tissue thickness and 1.4 kHz for 40% tissue thickness. Coherent compounding in the simulations downshifted these frequencies to about 0.50 kHz for both tissue regions. The frequencies with highest Fourier energy content in the biomechanics simulation were 0.93 kHz and 1.70 kHz for 15% and 40% tissue thickness respectively.

*3.3. Shear Wave Speed Analysis*

The quantitative analysis of shear wave observations consisted of shear modulus estimation based on group and phase velocity analysis for real and virtual SWE acquisitions, as visualized in Figure 6. For the measurements, the group velocity analysis provided median shear stiffness values of 14.6 kPa and 17.1 kPa for single and compounded PWI respectively, when discarding data of 5 edge elements for each shear wave during shear modulus estimation. These estimations increased to 23.8 kPa and 18.8 kPa when 15 more data points were not considered during the fitting procedure for each shear wave. Phase velocity analysis gave median values of 24.7 kPa and 27.3 kPa for the measurements. For single PWI, the stiffness range of TOF-estimations when taking less data points into account during fitting (15.9 kPa) was remarkably higher than for other stiffness estimation methods (5.0 kPa and 4.6 kPa for group and phase speed analysis respectively). Actual PVA stiffness was mechanically determined at $24.3 \pm 0.6$ kPa.

For the virtual imaging acquisitions, the group velocity-based method estimated median stiffness at 14.4 kPa and 15.4 kPa for single and compounded PWI respectively. When discarding data from 20 edge probe elements, TOF stiffness estimations increased to 18.0 kPa and 15.6 kPa. Phase velocity analysis provided, for both imaging simulations, higher estimates of median shear stiffness, i.e., 24.9 kPa and 25.2 kPa for single and compounded PWI respectively. As for the experiments, the largest spread in stiffness estimation across depth (11.8 kPa) was obtained for single PWI when applying the group velocity analysis and discarding data from 20 probe elements. For the biomechanics simulations, median shear stiffness of 16.5 kPa, 21.8 kPa and 24.7 kPa were obtained for group velocity (discarding 5 data points), group velocity (discarding 20 data points) and phase velocity analysis respectively. Again, the depth-dependency of stiffness estimations was the largest for the group speed method taking less data points into account during fitting (10.3 kPa).

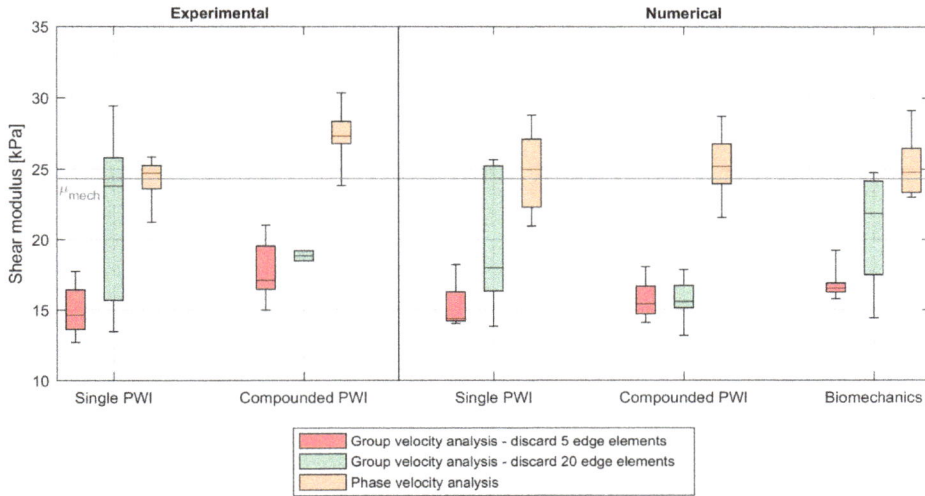

**Figure 6.** Comparison of estimated shear modulus via material characterization methods based on group and phase velocity for the numerical (biomechanics and ultrasound simulations with single and compounded plane wave imaging (PWI)) and experimental shear wave acquisitions (single and compounded PWI). The mechanically determined shear modulus $\mu_{mech}$ of 24.3 kPa is also indicated in this figure, corresponding to the modeled stiffness. The boxplot represents the variation in shear modulus estimation throughout depth ($n = 10$), where the box displays first, second (median) and third quartiles and the whiskers indicate minima and maxima.

## 4. Discussion

### 4.1. Multiphysics Modeling

In this work, a SWE multiphysics modeling approach incorporating the biomechanics and imaging physics of the shear wave propagation problem was presented, providing valuable insights into how the ultrafast US sequence and signal processing affects the true shear wave's characteristics in the time and frequency domain, and the subsequent shear modulus characterization. Furthermore, a modeling approach offers the benefits of full flexibility at the level of the tissue mechanics (tissue geometry, material properties and tissue surrounding) and ultrasound physics (ARF configuration, imaging settings and processing techniques). This approach was applied to a low-viscous pediatric ventricular phantom model, displaying clear shear wave dispersion as can be seen from the frequency-dependent phase velocity in the Fourier spectra and the split shear wave front in the temporal shear wave pattern for both experiment and biomechanics model [23]. The ventricular geometry was mainly the cause of the observed dispersion, as incorporating the measured viscoelastic material properties in the model did not significantly alter the shear wave characteristics (see [23] for details). A similar multiphysics approach has already been used by Palmeri et al. [40] to study jitter errors and displacement underestimation in unbounded media, also in combination with experiments. Another study [41] used these same tools to investigate how parameters related to shear wave excitation and tracking affected the quality of shear wave speed images. However, both studies mimicked a different elastography technique, called Acoustic Radiation Force Imaging (ARFI), which employs conventional line-by-line scanning instead of plane wave imaging to visualize the shear wave propagation.

In general, the multiphysics model was capable of reproducing the experimental results (see Figures 3–6), indicating that the simulated biomechanical ground truth is a good representation of the actual shear wave physics occurring in the PVA phantom. However, there were also some discrepancies in shear wave visualization and characterization. For the shear wave's characteristics in the time and frequency domain, we firstly noticed a different axial velocity magnitude (see Figures 3 and 4) and Fourier energy amplitude (see Table 3) as a result of scaling the time-averaged acoustic intensity to 1500 W/cm$^2$ when calculating the numerical ARF. For the virtual compounded acquisition, there was the additional effect of large shear wave travel in between ultrasound frames in combination with the presence of high relaxation velocities (blue in Figure 4) in the biomechanical simulation, indicating that compounding in the simulation reduces the downward velocities (red in Figure 4) more than in the experiment. Secondly, there were also differences in the temporal axial velocity pattern (e.g., larger relaxation peak at the center of the phantom for the simulations) and frequency spectra (e.g., more secondary mode excitation at 40% tissue thickness in the simulations). This can potentially be attributed to: (i) the manner of shear wave excitation in the model, i.e., applying a time-averaged body force and interface pressure instead of modeling the longitudinal wave propagation in the focused US beam, including reflection and attenuation, (ii) the difference in location of the actual and virtual SWE acquisitions, and (iii) the unknown experimental dead time between the pushing and imaging sequence. It should also be kept in mind that the beamforming process for experiment and simulation was performed with different infrastructure, i.e., the Aixplorer system and the NTNU in-house developed beamformer, respectively. Next to these dissimilarities in shear wave pattern in time and frequency, there were also inconsistencies in shear modulus estimation (Figure 6). These discrepancies are partly due to the same factors, as explained above, influencing shear wave propagation patterns and thus also stiffness characterization. Additionally, the simulations are noise-free, allowing more reliable shear stiffness estimates for every shear wave propagation path across depth compared to the experiments. Another potential cause explaining the stiffness discrepancy between experiment and simulation is a wrongly modeled material stiffness (based on uniaxial mechanical testing), as the mechanical properties of the PVA phantom could alter in the time difference between mechanical testing and SWE experiment.

### 4.2. Effect of Ultrafast Imaging on SWE in the Studied Left Ventricular Model

We studied the effect of ultrafast imaging on SWE by comparing shear wave visualization and characterization obtained from US and CSM simulations of our left ventricular phantom model. When analyzing the temporal shear wave patterns of all simulations in Figure 4, a clear broadening of the shear wave front and underestimation of axial velocities is noticeable for both imaging acquisitions. A similar negative velocity bias was also recently reported when using coherent plane wave compounding for Doppler imaging [42]. Furthermore, the plane wave compounded images revealed a shear wave pattern different than the single plane wave images: the split shear wave front, clearly visible in the single plane wave acquisition, was less observable in the compounded images (Figure 4). Furthermore, the experimental compounded images in Figure 3 showed a completely merged wave front instead of the split wave front as observed in the single plane wave images. Even though this observation was less clearly noticeable in the simulations (due to the presence of a larger relaxation peak in between the split wave front compared to the measurements, as mentioned in Section 4.1), the multiphysics model still demonstrated that these observed differences in temporal characteristics of the shear wave are mainly attributed to the chosen imaging parameters, as both virtual imaging acquisitions were derived from the same true mechanical wave propagation (see Figure 4). We also investigated the subsequent changes in the shear wave's frequency characteristics, which showed that the detected excited frequencies, amplitudes and modes did not necessarily correspond to the ones excited in the biomechanical model. Indeed, the biomechanical frequency spectra are solely dependent on the model characteristics and the ARF properties [43], whereas the imaged spectra are also affected by plane wave imaging, acting as a low-pass filter, and by image processing techniques such as pixel averaging and slow time up-sampling.

Next to this qualitative investigation, we also quantitatively studied the effect of ultrafast imaging on the performance of SWE by comparing the SWE-derived shear modulus for both US and CSM simulations (see Figure 6). This study showed that ultrafast imaging had mainly an effect on stiffness characterization through the group speed method: single and compounded PWI simulations led to median stiffness underestimations of −2.2 kPa (−4.0 kPa when discarding 20 data points) and −1.1 kPa (−6.2 kPa when discarding 20 data points) respectively compared to the SWE-derived stiffness estimates from the biomechanical simulations. Additionally, the results of the TOF method when discarding 20 edge elements were very depth-dependent for the single PWI simulation. This was also observed for the experiments in Figure 6. This large dissimilarity in depth-dependency of the stiffness estimates is due to a difference in meaning of the fitted linear relationship in the TOF method when discarding more or less data points for the single PWI acquisition. When 20 data points are discarded during the fitting procedure, the fitted linear relationship represents the true non-shifted shear wave position throughout time which varies a lot across depth, whereas it depicts an averaged shear wave position in time when only 5 data points are discarded (Figure S1). The latter corresponds to the TOF shear wave characterization with compounded PWI (as can be seen in Figure 6 and Figure S1), as the compounded images already visualize the averaged shear wave behavior. Nevertheless, the shear modulus estimates are depth-dependent for all applied material characterization methods, as can be seen in the spread of the boxplots in Figure 6. For the group velocity analysis (discarding 5 data points), this is mainly due to the difference in the shear wave propagation pattern at the upper and lower boundaries of the phantom (±0–25% and ±75–100% depth) compared to the middle segment of the phantom (±25–75% depth), as visible in Figure S1. This group speed-derived stiffness difference between the boundaries and center of a tissue-mimicking medium was experimentally studied by Mercado et al. [44], in which they identified the presence of Scholte surface waves at the fluid–solid interface as the primary reason for this discrepancy. For the phase velocity analysis, the cause of the depth-dependency of the stiffness estimates is less straightforward, as the extracted frequency characteristics of the primary mode across depth were very similar (see Figure 5). However, as also shown in [23], characterizing deeper shear waves via the phase velocity analysis is more challenging as their 2D FFT energy content is smaller (fewer data points to fit) and their velocity amplitude is lower (lower signal-to-noise ratio), leading to less reliable shear modulus estimates.

Phase velocity analysis provided a more robust and correct estimate for both the biomechanics and imaging simulations, as spectral characteristics of the tracked primary mode (fitted to the theoretical A0-mode) for all acquisitions are very similar, as shown in Figure 5. Furthermore, for both experiment and simulation, the true tissue stiffness was underestimated by the TOF method, independent of the number of considered data points, whereas phase velocity analysis provided a better estimate of the mechanically determined stiffness. This is in accordance with our previous findings of experimental work on the same ventricular model in which we only applied single PWI [23]. Nevertheless, if the stiffness estimation technique is chosen based on observed shear wave physics (i.e., TOF method for compounded images visualizing almost no dispersion and phase velocity analysis for single plane wave images depicting dispersion), differences of minimally 5.9 kPa and 9.3 kPa are obtained for measurements and simulations, respectively. This is about 25% of the value of the actual shear modulus, and non-negligible. Therefore, when studying low viscous settings evoking guided wave dispersion due to geometry, one should be cautious when selecting a tissue characterization method based on the observed shear wave pattern as this might be affected by the applied imaging set-up. In these cases, it might be relevant to also study phase velocity next to group velocity.

It should be noted that the primary objective of this work was not to compare the performance of single and compounded PWI, as this requires (i) the study of multiple configurations and material models, (ii) the use of more complex SWE-based material characterization and (iii) the inclusion of noise in the numerical models. However, this work shows the potential of computational modeling in identifying potential pitfalls in shear wave visualization and characterization with SWE, demonstrated through a case study of an idealized SWE setting with little amount of noise (as shown by the good correspondence between experiment and simulation). Future research should focus on applying the current modeling technique to different settings to further study the performance of single and compounded PWI.

### 4.3. Recommendations and Impact for Other Applications

The dispersive shear wave propagation pattern studied here is inherently linked to the considered setting, i.e., a left ventricular low viscous phantom with pediatric geometry. We focused on the isolated effect of guided wave dispersion due to geometry, and therefore, the formulated conclusions cannot simply be extrapolated to actual tissue settings as dispersion in tissues can be caused by a combination of varying factors such as geometry, viscosity and non-homogeneous (potentially anisotropic) material characteristics. This is among other things noticeable in the excited frequency range of the studied shear wave (up to 2 kHz), which is much larger than the conventional 1 kHz shear wave frequency spectra reported in real tissue settings due to tissue's high shear viscosity [17]. Additionally, the observed shear wave fronts were quite isotropic in all directions of the shear wave paths in 2D, whereas these will become guided along the fiber orientation in anisotropic tissue [45,46]. These true tissue characteristics demand more advanced tissue characterization algorithms as now (i) an isotropic bulky elastic material is assumed in the group speed analysis in order to apply Formula (4), and (ii) a theoretical dispersion curve of an isotropic homogeneous elastic plate in water is used as fitting ground truth in the phase speed analysis. Therefore, complementary research is necessary to investigate how the formulated conclusions concerning shear wave visualization and characterization are translated to actual tissue settings in vivo, particularly when assessing the effect of compounding.

Despite these dissimilarities between shear wave physics in the phantom-model and actual tissue, the multiphysics model of the presented case study allowed the assessment of the effect of ultrafast imaging on shear wave visualization and characterization from a mechanical point of view, as described in the previous section. Furthermore, this study showed that the number of compounding angles (i.e., the factor with which the frame rate is reduced) should be chosen taking the maximal reachable PRF (linked to imaged depth and technical capabilities of the ultrasound system), the wave propagation speed of the investigated material (related to its mechanical properties) and the bandwidth of the imaged phenomenon (related to different absorption mechanisms such as viscosity) into account. The resulting compounded frame rate should be sufficiently high to obtain an accurate representation of the mechanical shear wave physics, which was not the case for the studied left ventricular phantom model. Additionally, a high frame rate is also desirable from the shear wave characterization point of view, as this means a high Nyquist cut-off frequency, providing a more extensive Fourier spectrum, and thus a more reliable stiffness estimate via the phase velocity analysis.

Similar recommendations were recently published by Widman et al. [19], who studied the optimal ARF and imaging settings to maximize bandwidth for phase velocity analysis in SWE on ex vivo arterial settings. In their study on arterial stiffness estimation, they claimed that a high PRF with poorer image quality is more desirable than a lower PRF with better image quality.

### 5. Conclusions

In this work, we assessed the effect of ultrafast imaging on dispersive shear wave visualization and subsequent shear stiffness characterization by means of SWE experiments in combination with a multiphysics model of a LV phantom model with pediatric geometry. This model offers the advantage of giving access to the true biomechanical wave propagation, which is unknown in the SWE measurements. The multiphysics model of the idealized LV phantom revealed that the detected shear wave features in the time and frequency domain by ultrafast imaging do not necessarily depict the ARF-excited characteristics of the biomechanical model. Furthermore, application of the compounding technique in ultrafast imaging even altered the dispersion features in the temporal shear wave pattern for both experiments and simulations, leading to a stiffness underestimation of minimally 25% when choosing a group velocity-based algorithm instead of a phase velocity one. Additionally, the applied group speed material characterization method was very sensitive to the applied algorithm settings (such as the number of tracked data points) and the selected axial depth, as ultrafast imaging can alter the shear wave front location in the shear wave visualization. Therefore, it is important to keep a high frame rate during compounding in order to obtain an accurate representation of shear wave physics and the subsequently derived material stiffness. Future research should focus on investigating additional configurations with more advanced SWE-based material characterization to further generalize these conclusions. Nevertheless, this work presents a versatile and powerful simulation environment to evaluate the performance of ultrafast imaging in shear wave visualization and characterization with SWE, and to identify potential pitfalls in accurately capturing shear wave propagation.

**Supplementary Materials:** The following are available online at http://www.mdpi.com/2076-3417/7/8/840/s1. Figure S1: Illustration of the effect of discarding 5 or 20 data points at the edges of each shear wave during the fitting procedure in the Time Of Flight (TOF) method: a comparison between different depths and imaging acquisitions.

**Acknowledgments:** Annette Caenen is the recipient of a research grant from the Flemish government agency for Innovation and Entrepreneurship (VLAIO). Darya Shcherbakova is supported by the Research Foundation Flanders (FWO). This research was also supported by a grant for international mobility from the CWO (committee for scientific research at the Faculty of Engineering and Architecture at Ghent University) to facilitate the research visits at the Institut Langevin (Paris, France).

**Author Contributions:** Annette Caenen, Luc Mertens, Abigail Swillens and Patrick Segers conceived and designed the experiments/simulations; Annette Caenen and Darya Shcherbakova performed the experiments under supervision of Mathieu Pernot; Annette Caenen analyzed the experimental data; Annette Caenen performed the finite element simulations; Annette Caenen performed ultrasound simulations with help from Abigail Swillens and Ingvild Kinn Ekroll; Annette Caenen is the main author of the paper; critical revision were provided by all co-authors.

## References

1. Bercoff, J. Ultrafast Ultrasound Imaging. In *Ultrasound Imaging—Medical Applications*; Minin, P.O., Ed.; Intech: Rijeka, Croatia, 2011.

2. Tanter, M.; Fink, M. Ultrafast imaging in biomedical ultrasound. *IEEE Trans. Ultrason. Ferroelectr. Freq. Control* **2014**, *61*, 102–119. [CrossRef] [PubMed]

3. Tanter, M.; Bercoff, J.; Sandrin, L.; Fink, M. Ultrafast Compound Imaging for 2-D Motion Vector Estimation: Application to Transient Elastography. *IEEE Trans. Ultrason. Ferroelectr. Freq. Control* **2002**, *49*, 1363–1374. [CrossRef] [PubMed]

4. Sandrin, L.; Tanter, M.; Catheline, S.; Fink, M. Shear Modulus Imaging with 2-D Transient Elastography. *IEEE Trans. Ultrason. Ferroelectr. Freq. Control* **2002**, *49*, 426–435. [CrossRef] [PubMed]

5. Bercoff, J.; Tanter, M.; Fink, M. Supersonic Shear Imaging: A New Technique for Soft Tissue Elasticity Mapping. *IEEE Trans. Ultrason. Ferroelectr. Freq. Control* **2004**, *51*, 396–409. [CrossRef] [PubMed]

6. Gennisson, J.L.; Deffieux, T.; Fink, M.; Tanter, M. Ultrasound elastography: Principles and techniques. *Diagn. Interv. Imaging* **2013**, *94*, 487–495. [CrossRef] [PubMed]

7. Sarvazyan, A.P.; Rudenko, O.V.; Swanson, S.D.; Fowlkes, J.B.; Emelianov, S.Y. Shear Wave Elasticity Imaging: A New Ultrasonics Technology of Medical Diagnostics. *Ultrasound Med. Biol.* **1998**, *24*, 1419–1435. [CrossRef]

8. Nightingale, K.; Soo, M.S.; Nightingale, R.; Trahey, G. Acoustic Radiation Force Impulse Imaging: In Vivo Demonstration of Clinical Feasibility. *Ultrasound Med. Biol.* **2002**, *28*, 227–235. [CrossRef]

9. Doherty, J.R.; Trahey, G.E.; Nightingale, K.R.; Palmeri, M.L. Acoustic radiation force elasticity imaging in diagnostic ultrasound. *IEEE Trans. Ultrason. Ferroelectr. Freq. Control* **2013**, *60*, 685–701. [CrossRef] [PubMed]

10. Montaldo, G.; Tanter, M.; Bercoff, J.; Benech, N.; Fink, M. Coherent Plane-Wave Compounding for Very High Frame Rate Ultrasonography and Transient Elastography. *IEEE Trans. Ultrason. Ferroelectr. Freq. Control* **2009**, *56*, 489–506. [CrossRef] [PubMed]

11. Bercoff, J.; Montaldo, G.; Loupas, T.; Savery, D.; Mézière, F.; Fink, M.; Tanter, M. Ultrafast Compound Doppler Imaging: Providing Full Blood Flow Characterization. *IEEE Trans. Ultrason. Ferroelectr. Freq. Control* **2011**, *58*, 134–147. [CrossRef] [PubMed]

12. Papadacci, C.; Pernot, M.; Couade, M.; Fink, M.; Tanter, M. High Contrast Ultrafast Imaging of the Human Heart. *IEEE Trans. Ultrason. Ferroelectr. Freq. Control* **2014**, *61*, 288–301. [CrossRef] [PubMed]

13. Couture, O.; Bannouf, S.; Montaldo, G.; Aubry, J.F.; Fink, M.; Tanter, M. Ultrafast imaging of ultrasound contrast agents. *Ultrasound Med. Biol.* **2009**, *35*, 1908–1916. [CrossRef] [PubMed]

14. Mace, E.; Montaldo, G.; Osmanski, B.F.; Cohen, I.; Fink, M.; Tanter, M. Functional ultrasound imaging of the brain: Theory and basic principles. *IEEE Trans. Ultrason. Ferroelectr. Freq. Control* **2013**, *60*, 492–506. [CrossRef] [PubMed]

15. Tanter, M.; Bercoff, J.; Athanasiou, A.; Deffieux, T.; Gennisson, J.L.; Montaldo, G.; Muller, M.; Tardivon, A.; Fink, M. Quantitative assessment of breast lesion viscoelasticity: Initial clinical results using supersonic shear imaging. *Ultrasound Med. Biol.* **2008**, *34*, 1373–1386. [CrossRef] [PubMed]

16. Ferraioli, G.; Parekh, P.; Levitov, A.B.; Filice, C. Shear wave elastography for evaluation of liver fibrosis. *J. Ultrasound Med.* **2014**, *33*, 197–203. [CrossRef] [PubMed]

17. Couade, M.; Pernot, M.; Prada, C.; Messas, E.; Emmerich, J.; Bruneval, P.; Criton, A.; Fink, M.; Tanter, M. Quantitative assessment of arterial wall biomechanical properties using shear wave imaging. *Ultrasound Med. Biol.* **2010**, *36*, 1662–1676. [CrossRef] [PubMed]

18. Nguyen, T.M.; Couade, M.; Bercoff, J.; Tanter, M. Assessment of Viscous and Elastic Properties of Sub-Wavelength Layered Soft Tissues Using Shear Wave Spectroscopy: Theoretical Framework and In Vitro Experimental Validation. *IEEE Trans. Ultrason. Ferroelectr. Freq. Control* **2011**, *58*, 2305–2315. [CrossRef] [PubMed]

19. Widman, E.; Maksuti, E.; Amador, C.; Urban, M.W.; Caidahl, K.; Larsson, M. Shear Wave Elastography Quantifies Stiffness in Ex Vivo Porcine Artery with Stiffened Arterial Region. *Ultrasound Med. Biol.* **2016**, *42*, 2423–2435. [CrossRef] [PubMed]

20. Caenen, A.; Shcherbakova, D.; Verhegghe, B.; Papadacci, C.; Pernot, M.; Segers, P.; Swillens, A. A Versatile and Experimentally Validated Finite Element Model to Assess the Accuracy of Shear Wave Elastography in a Bounded Viscoelastic Medium. *IEEE Trans. Ultrason. Ferroelectr. Freq. Control* **2015**, *62*, 439–450. [CrossRef] [PubMed]

21. Palmeri, M.L.; Sharma, A.C.; Bouchard, R.R.; Nightingale, R.W.; Nightingale, K.R. A Finite-Element Method Model of Soft Tissue Response to Impulsive Acoustic Radiation Force. *IEEE Trans. Ultrason Ferroelectr. Freq. Control* **2005**, *52*, 1699–1712. [CrossRef] [PubMed]

22. Lee, K.H.; Szajewski, B.A.; Hah, Z.; Parker, K.J.; Maniatty, A.M. Modeling shear waves through a viscoelastic medium induced by acoustic radiation force. *Int. J. Numer. Methods Biomed. Eng.* **2012**, *28*, 678–696. [CrossRef] [PubMed]

23. Caenen, A.; Pernot, M.; Shcherbakova, D.A.; Mertens, L.; Kersemans, M.; Segers, P.; Swillens, A. Investigating Shear Wave Physics in a Generic Pediatric Left Ventricular Model via In Vitro experiments and Finite Element Simulations. *IEEE Trans. Ultrason. Ferroelectr. Freq. Control* **2017**, *64*, 349–361. [CrossRef] [PubMed]

24. Jensen, J.A. Field: A Program for Simulating Ultrasound Systems. In Proceedings of the 10th Nordic-Baltic Conference on Biomedical Imaging, Tampere, Finland, 9–13 June 1996; pp. 351–353.

25. Jensen, J.A.; Svensen, N.B. Calculation of Pressure Fields from Arbitrarily Shaped, Apodized, and Excited Ultrasound Transducers. *IEEE Trans. Ultrason. Ferroelectr. Freq. Control* **1992**, *39*, 262–267. [CrossRef] [PubMed]

26. Nightingale, K. Acoustic Radiation Force Impulse (ARFI) Imaging: A Review. *Curr. Med. Imaging Rev.* **2011**, *7*, 328–339. [CrossRef] [PubMed]

27. Shutilov, V. *Fundamental Physics of Ultrasound*; CRC Press: Boca Raton, FL, USA, 1988.

28. *Standard Test Methods for Density and Specific Gravity (Relative Density) of Plastics by Displacement*; ASTM Standard D792-08; ASTM International: West Conshohocken, PA, USA, 2008.

29. Kamopp, D.C.; Margolis, D.L.; Rosenberg, R.C. Appendix: Typical material property values useful in modeling mechanical, acoustic and hydraulic elements. In *System Dynamics: Modeling, Simulation and Control of Mechatronic Systems*; John Wiley & Sons, Inc.: Hoboken, NJ, USA, 2012.

30. Dendy, P.; Heaton, B. *Physics for Diagnostic Radiology*; CRC Press: Boca Raton, FL, USA, 2012.

31. Schroeder, W.; Martin, K.; Lorensen, B. *The Visualization Toolkit*, 4th ed.; Kitware Inc.: Clifton Park, NY, USA, 2006.

32. Shcherbakova, D. A multiphysics model of the mouse aorta for the mice optimization of high-frequency ultrasonic imaging in mice. In *Faculty of Engineering and Architecture*; Ghent University: Ghent, Belgium, 2012.

33. Ekroll, I.K.; Swillens, A.; Segers, P.; Dahl, T.; Torp, H.; Lovstakken, L. Simultaneous quantification of flow and tissue velocities based on multi-angle plane wave imaging. *IEEE Trans. Ultrason. Ferroelectr. Freq. Control* **2013**, *60*, 727–738. [CrossRef] [PubMed]

34. Kasai, C.; Namekawa, K.; Koyano, A.; Omoto, R. Real-Time Two-Dimensional Blood Flow Imaging Using an Autocorrelation Technique. *IEEE Trans. Sonics Ultrason.* **1985**, *32*, 458–464. [CrossRef]

35. Lovstakken, L. *Signal Processing in Diagnostic Ultrasound: Algorithms for Real-Time Estimation and Visualization of Blood Flow Velocity*; Norwegian University of Science and Technology: Trondheim, Norway, 2007.

36. Palmeri, M.L.; Wang, M.H.; Dahl, J.J.; Frinkley, K.D.; Nightingale, K.R. Quantifying hepatic shear modulus in vivo using acoustic radiation force. *Ultrasound Med. Biol.* **2008**, *34*, 546–558. [CrossRef] [PubMed]

37. Bernal, M.; Nenadic, I.; Urban, M.W.; Greenleaf, J.F. Material property estimation for tubes and arteries using ultrasound radiation force and analysis of propagating modes. *J. Acoust. Soc. Am.* **2011**, *129*, 1344–1354. [CrossRef] [PubMed]

38. Nenadic, I.Z.; Urban, M.W.; Mitchell, S.A.; Greenleaf, J.F. Lamb wave dispersion ultrasound vibrometry (LDUV) method for quantifying mechanical properties of viscoelastic solids. *Phys. Med. Biol.* **2011**, *56*, 2245–2264. [CrossRef] [PubMed]

39. Kanai, H. Propagation of spontaneously actuated pulsive vibration in human heart wall and in vivo viscoelasticity estimation. *IEEE Trans. Ultrason. Ferroelectr. Freq. Control* **2005**, *52*, 1931–1942. [CrossRef] [PubMed]

40. Palmeri, M.L.; McAleavey, S.A.; Trahey, G.E.; Nightingale, K.R. Ultrasonic Tracking of Acoustic Radiation Force-Induced Displacements in Homogeneous Media. *IEEE Trans. Ultrason. Ferroelectr. Freq. Control* **2006**, *53*, 1300–1313. [CrossRef] [PubMed]

41. Rouze, N.C.; Wang, M.H.; Palmeri, M.L.; Nightingale, K.R. Parameters affecting the resolution and accuracy of 2-D quantitative shear wave images. *IEEE Trans. Ultrason. Ferroelectr. Freq. Control* **2012**, *59*, 1729–1740. [CrossRef] [PubMed]

42. Ekroll, I.K.; Voormolen, M.M.; Standal, O.K.V.; Rau, J.M.; Lovstakken, L. Coherent compounding in doppler imaging. *IEEE Trans. Ultrason. Ferroelectr. Freq. Control* **2015**, *62*, 1634–1643. [CrossRef] [PubMed]

43. Palmeri, M.L.; Deng, Y.; Rouze, N.C.; Nightingale, K.R. Dependence of shear wave spectral content on acoustic radiation force excitation duration and spatial beamwidth. *IEEE Trans. Ultrason. Ferroelectr. Freq. Control* **2014**. [CrossRef]

44. Mercado, K.P.; Langdon, J.; Helguera, M.; McAleavey, S.A.; Hocking, D.C.; Dalecki, D. Scholte wave generation during single tracking location shear wave elasticity imaging of engineered tissues. *J. Acoust. Soc. Am.* **2015**, *138*, EL138–EL144. [CrossRef] [PubMed]

45. Couade, M.; Pernot, M.; Messas, E.; Bel, A.; Ba, M.; Hagege, A.; Fink, M.; Tanter, M. In vivo quantitative mapping of myocardial stiffening and transmural anisotropy during the cardiac cycle. *IEEE Trans. Med. Imaging* **2011**, *30*, 295–305. [CrossRef] [PubMed]

46. Lee, W.N.; Pernot, M.; Couade, M.; Messas, E.; Bruneval, P.; Bel, A.; Hagege, A.; Fink, M.; Tanter, M. Mapping myocardial fiber orientation using echocardiography-based shear wave imaging. *IEEE Trans. Med. Imaging* **2012**, *31*, 554–562. [PubMed]

*applied*
*sciences*

MDPI

Article

# Automatic Definition of an Anatomic Field of View for Volumetric Cardiac Motion Estimation at High Temporal Resolution

Alejandra Ortega [1,*], João Pedrosa [1], Brecht Heyde [1], Ling Tong [1,2] and Jan D'hooge [1]

[1] Department of Cardiovascular Sciences, KU Leuven, 3000 Leuven, Belgium; joao.pedrosa@kuleuven.be (J.P.); brecht.jk.heyde@gmail.com (B.H.); tonglingpku@gmail.com (L.T.); jan.dhooge@uzleuven.be (J.D.)
[2] Department of Biomedical Engineering, Tsinghua University, 100084 Beijing, China
[*] Correspondence: alejao16@gmail.com

Received: 7 June 2017; Accepted: 20 July 2017; Published: 24 July 2017

**Abstract:** Fast volumetric cardiac imaging requires reducing the number of transmit events within a single volume. One way of achieving this is by limiting the field of view (FOV) of the recording to the myocardium when investigating cardiac mechanics. Although fully automatic solutions towards myocardial segmentation exist, translating that information in a fast ultrasound scan sequence is not trivial. In particular, multi-line transmit (MLT) scan sequences were investigated given their proven capability to increase frame rate (FR) while preserving image quality. The aim of this study was therefore to develop a methodology to automatically identify the anatomically relevant conically shaped FOV, and to translate this to the best associated MLT sequence. This approach was tested on 27 datasets leading to a conical scan with a mean opening angle of $19.7° \pm 8.5°$, while the mean "thickness" of the cone was $19° \pm 3.4°$, resulting in a frame rate gain of about 2. Then, to subsequently scan this conical volume, several MLT setups were tested in silico. The method of choice was a 10MLT sequence as it resulted in the highest frame rate gain while maintaining an acceptable cross-talk level. When combining this MLT scan sequence with at least four parallel receive beams, a total frame rate gain with a factor of approximately 80 could be obtained. As such, anatomical scan sequences can increase frame rate significantly while maintaining information of the relevant structures for functional myocardial imaging.

**Keywords:** fast 3D cardiac imaging; anatomical imaging; multi-line transmit

## 1. Introduction

Over the last decades, volumetric cardiac ultrasound imaging has gained momentum as the modality of choice to assess cardiac morphology and visualize global heart motion [1]. Recently, 3D cardiac ultrasound has also been used to quantitatively assess regional cardiac dynamics and several commercial products for 3D speckle tracking are now readily available [2]. One of the challenges for 3D motion estimation remains the relatively low image quality of the volumetric ultrasound data set. Particularly, its relatively low spatiotemporal resolution is of concern. Indeed, state-of-the-art commercial systems make use of a combination of several techniques in order to improve frame rate while maintaining image quality, such as limiting the field of view (FOV) [3], decreasing the line density [4], using electrocardiography (ECG) gating [5], and/or applying parallel receive beamforming (i.e., multi-line acquisition, MLA) [6]. This typically results in a frame rate of up to ~30 Hz when scanning with a moderate line density, when using a representative wide-angle field of view (i.e., ~60° × ~ 60°) and when gating is performed over four to six cardiac cycles. ECG gating not only lengthens the acquisition but can also induce artefacts due to arrhythmias, breathing, motion or incorrect gating. A frame rate of 30 Hz for a 60° × 60° opening angle is adequate to evaluate cardiac morphology and visualize global heart motion [1]. However, when trying to visualize the mechanical activation wave associated with the contraction of the heart, a frame rate around 500 Hz is needed [4], while assessing the mechanical properties of the heart by measuring the speed by which (acoustically induced) shear waves propagate through the myocardium requires even higher frame rates [7]. Based on the desired cardiac application, the temporal resolution of the imaging system should thus be adapted.

Recently, two other solutions have been proposed to increase the temporal resolution of the 3D systems. On the one hand, Diverging Wave Imaging (DWI) has been introduced, which uses a sparse virtual array located behind the probe [8]. This allows volumetric imaging at very high rates since the line density is only restricted by the reconstruction time. However, the use of unfocused beams implies that the energy of the beam is spread over a wider area resulting in low pressures which do not allow for harmonic imaging. In addition, the overall signal-to-noise ratio drops, having a negative impact on motion estimators (as illustrated by the Cramer–Rao Lower Bound [9]).

On the other hand, multi-line transmit beamforming (MLT) has been proposed to increase temporal resolution by simultaneously transmitting multiple focused beams allowing harmonic imaging [10]. While its frame rate gain is more limited than that of diverging wave imaging [11] and although potential artifacts such as cross-talk might be introduced due to the interaction between neighboring MLT beams, it preserves image quality [10] and is an attractive imaging approach. Indeed, it has recently been demonstrated that cross-talk artifacts can be suppressed by using proper apodization [11], by avoiding the main directions of the transducer [6], by frequency-coding the different MLT beams [12,13], and/or by physically separating the beams in space [9]. Furthermore, MLT can easily be combined with MLA to further increase the temporal resolution. More precisely, the width of the transmit beams can be customized based on the desired amount of receive parallelization for each transmit beam. Based on Expression (10) of [14], at least 4MLA (two in azimuth and two in elevation) can be reconstructed per focused beam in 3D for a typical cardiac 2D matrix array transducer. In order to allow a higher number of MLA, broader beams can be transmitted by reducing the transmit aperture. It should also be noted that safety issues due to the acoustic superposition of the MLT beams in the near field are of little concern. Indeed, we recently demonstrated that by small modifications of the transmit beamforming (i.e., introducing small delays or phase shifts between different MLT beams), the near-field pressure can be adjusted to remain within safety regulations [15]. Finally, in order to test the feasibility of MLT for volumetric imaging, a qualitative [16] and a quantitative [17] study were recently performed where the 2D findings were extrapolated to 3D. It was shown, by in silico and in vitro experiments, that a 16MLT-4MLA with transmit and receive apodization (Tukey $\alpha = 0.5$) generates volumetric images within a single heartbeat with an appropriate image quality for functional cardiac imaging [17].

Despite these fast imaging solutions, it is important to note that the typical volumetric ultrasound recording remains pyramidal, implying that a significant portion of the reconstructed image lines are not relevant to analyze 3D myocardial dynamics. Therefore, a straightforward way to reduce the number of transmit events is by limiting the FOV to the anatomically relevant space only, i.e., to a conically shaped volume, which captures the myocardium throughout the cardiac cycle. This reduced FOV can then be combined with the aforementioned parallelized scan sequences in order to further increase frame rate.

Although fully automatic solutions towards myocardial segmentation exist, translating that information in a fast ultrasound scan sequence is not obvious. The aim of this study was therefore to (1) develop a methodology to automatically define the characteristics of the conically shaped FOV given a segmented myocardial volume, and (2) to set up an appropriate MLT scan sequence to scan the associated FOV as fast as possible. Ideally, the FOV definition and its associated parallel scan sequence would be implemented in real-time and customized per patient.

## 2. Methods

### 2.1. Anatomical Relevant Space

The relevant FOV for functional cardiac imaging can be defined from a 3D dataset as follows:

1. Automatic real-time segmentation of myocardial boundaries

   A fully automatic real-time segmentation of the left ventricular myocardium in a volumetric ultrasound recording was performed using the B-spline Explicit Active Surfaces (BEAS) framework [18,19]. More specifically, BEAS uses two explicit functions, one to represent the endocardial surface and another to represent the myocardial thickness. This allows to fully characterize the endo- and epicardial surfaces. These surfaces can then be used to define a binary mask identifying voxels belonging to the myocardium only.

2. Coverage function

   Using these binary images, a "coverage function" was defined as follows. First, based on the ray-tracing principle, the path of a given scanline within the volumetric image volume can be traced. The pixels belonging to that scanline are compared with the binary mask, in order to compute the percentage of pixels of the given scanline belonging to the myocardium. Finally, this procedure is repeated for all scan lines in the original pyramidal volume leading to a "coverage function".

3. Ring-shaped template matching

   To find a spatially continuous FOV that covers a given percentage of the total amount of myocardium (i.e., prospectively defined by the user as "$T$"), a ring-shaped template matching was used. This shape was chosen as an approximation of the left ventricular geometry when looking down from the apex, i.e., when the transducer is placed in an apical position. In 3D, this FOV therefore defines a hollowed cone, Figure 1 (left). We express the amount of myocardial coverage $T$ as a function of the inner radius of the ring template ($R_i$) and its thickness ($\Delta R$). In order to effectively gain frame rate, a compromise has to be made between the amount of myocardial coverage, i.e., $T$, and the extent of the FOV. From all $R_i$ and $\Delta R$ combinations that provide $T$ myocardial coverage, the one with minimal $\Delta R$ was chosen, as this would keep the volume to be scanned minimal. In this way, it is ensured that the desired $T$ coverage is obtained using the least amount of lines possible (i.e., at the highest frame rate). In turn, these radii are used to determine the parameters (opening angle, $\Phi$, and thickness, $d\Phi$) for a conical scan, as represented in Figure 1 (right). These radii give directly the inner and outer image lines that define the inner and outer surface of the cone. Then, using the angular inter-beam spacing, the line numbers can be converted to the respective angles.

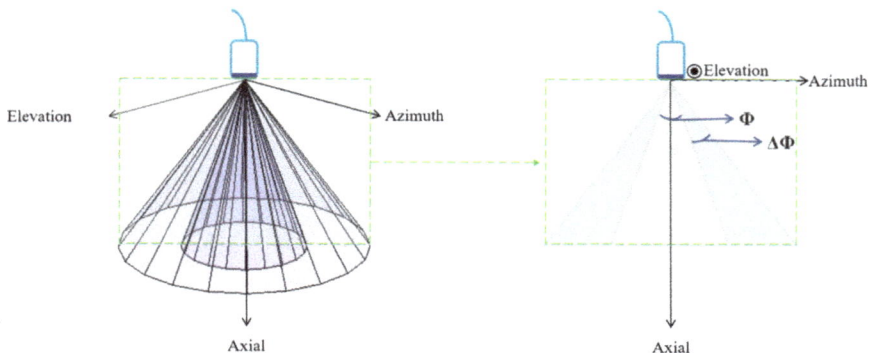

**Figure 1.** Graphical representation of a conical scan for anatomical imaging in 3D (**left**) and 2D (**right**), where $\Phi$ and $\Delta\Phi$ represent the opening angle and thickness of the cone, respectively.

## 2.2. Parallelized Scan Sequence

In order to find the best scan sequence for a given FOV, the cross-talk energy level of several MLT setups can be quantified by computer simulation. In particular, a 2D square phased array transducer (10 × 10 mm$^2$, 32 × 32 elements, 312.5 μm pitch) transmitting a Gaussian pulse with a central frequency of 2.5 MHz and 50% bandwidth was simulated using a graphical computer unit (GPU) based implementation of the spatial impulse response method (Simpulse) [20]. The transmit beams were focused at a depth of 60 mm while dynamic focusing was used during receive. Tukey apodization (α = 0.5) was applied in both transmit and receive to suppress cross-talk artifacts. Furthermore, the MLT beams were spread circumferentially around the conically shaped FOV in an equiangular manner, and were staggered in the radial direction between neighboring beams to increase the inter-beam spacing and therefore to reduce cross-talk. Such staggering was chosen to be half the thickness of the hollow cone at the focal depth. The cross-talk energy level of the MLT systems can be determined using the process described in [17]. As such, simulations of the two-way beam profiles were performed on a plane parallel to the transducer surface (i.e., C-plane) at a depth of 65 mm (i.e., around halfway the common imaging depth for cardiac applications). This depth was chosen based on our previous in silico findings [17] while using the exact same system configuration. It was shown that the best performing system at this depth was also the system of choice at other depths and steering angles. In this study, we therefore focused on evaluating the systems' performance at this depth only.

To quantify how the energy of a beam spreads in the volume and potentially interferes with other MLT beams, the same procedure as in [17] was followed. In short, a beam was transmitted into one of the MLT directions ($\theta_i$) and subsequently received along each of the MLT beam directions (including the same transmit direction: $\theta_{1...i...j}$). Next, the sum of the energy while transmitting and receiving in the same MLT direction was considered as single-line-transmission (SLT) energy (i.e., the ideal situation), whereas cross-talk was computed as the sum of the energy that deviates from the transmit beam towards the other remaining MLT directions, i.e., when transmitting in a given MLT direction but receiving in one of the other remaining MLT directions. This procedure was repeated for every MLT beam in the setup. The final cross-talk energy level ($E_{xtalk}$) was quantified in decibels. Mathematically, this can be expressed as:

$$E_{xtalk} = 10 \cdot log10 \left( \frac{\sum_{i=1}^{\#mlt} \sum_{j=1, j \neq i}^{\#mlt} E^{\theta_{ij}}}{\sum_{i=1}^{\#mlt} E^{\theta_{ii}}} \right) \quad (1)$$

where: $i$ is a given MLT direction being assessed; $j$ is a MLT direction different from the direction being tested ($i$); $\#mlt$ is the total number of MLT beams in the system; $E^{\theta_{ij}}$ is the energy in the C-plane when transmitting in a given MLT direction $i$ and receiving in one of the other directions ($j$) (i.e., cross-talk); and, $E^{\theta_{ii}}$ is the energy in the C-plane when transmitting in a given MLT direction $i$ while receiving in the same direction $i$ (i.e., SLT energy).

Please note that in the present study, the FOV definition and its associated parallel scanning sequence have been proposed considering apical view acquisitions only.

## 3. Experiments

To test this approach, 27 volumetric ultrasound datasets were randomly selected from the DOPPLER-CIP database (a large multi-center clinical FP7-funded study), which targeted patients suspected of (chronic) ischemia [21]. The mean opening angle for the volumetric scans in azimuth and elevation was 53.5°, resulting in approximately 7282 scanlines. The data was acquired using a GE E9 scanner (GE Vingmed, Horten, Norway) equipped with a 4 V transducer. The segmentation was performed at the end-diastole. A coverage threshold $T$ of 85% was chosen.

After finding the anatomical relevant space, four systems were investigated: 8MLT, 9MLT, 10MLT and 12MLT as illustrated in Figure 2. The cross-talk energy level of the systems was quantified using Equation (1), and a threshold of −30 dB was defined to give acceptable B-mode signal-to-noise ratio (SNR). Although this pre-defined cut-off was arbitrarily chosen from a retrospective analysis of our

2D in silico and in vivo findings, 3D in vitro experiments have nevertheless corroborated that this is a suitable measurement, even maybe a bit too restrictive, to ensure an adequate image quality [13].

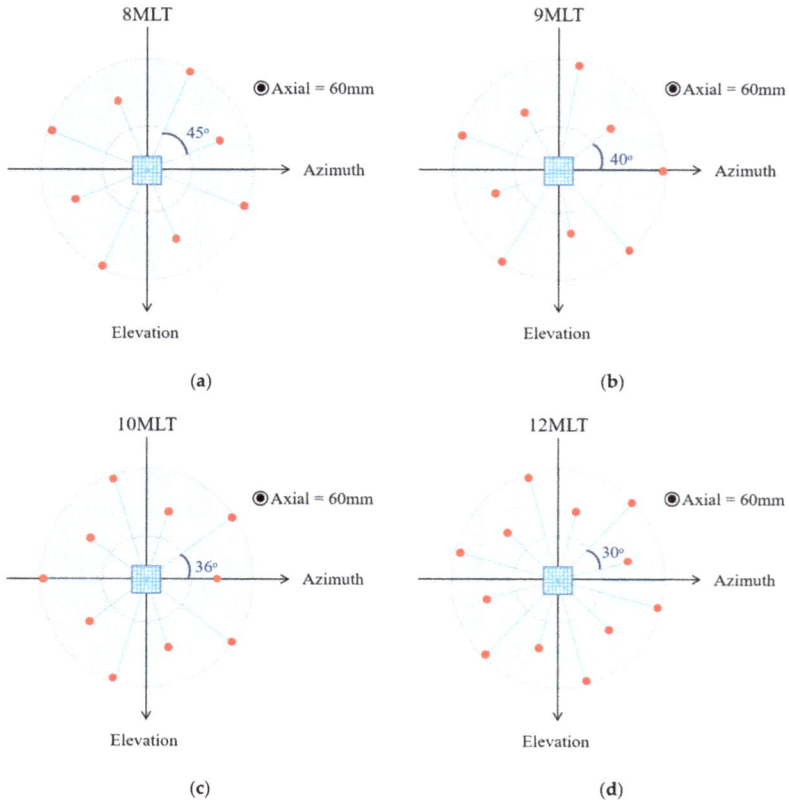

**Figure 2.** View of the tested multi-line transmit (MLT) systems looked from a C-plane at 60 mm with respect to the rectangular aperture (central blue square). The MLT beams are spread out in an equiangular manner in the circumferential direction of the cone while steered in the radial direction. In this way, the beams are as far apart as possible and therefore the inter-beam interference is reduced.

## 4. Results

An example of the workflow to determine the anatomical relevant space is shown in Figure 3. Figure 3a illustrates the segmented myocardial boundaries as overlays over the three orthogonal slices through the middle of the ultrasonic volume. Given these binary coordinates, the myocardial coverage per line is quantified as displayed in Figure 3b. As expected, the lines crossing the ventricular walls have the highest percentage of coverage while the ones crossing the apex have the lowest. Finally, the percentage of coverage can be chosen using the ring-shaped template matching. Logically, the smaller the inner radius and the bigger the outer radius, the higher the total myocardial coverage, Figure 3c.

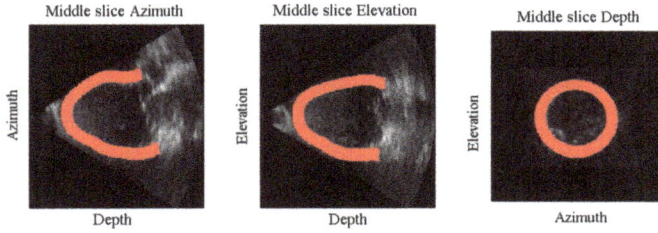

Patient with cardiac ischemia suspicion - Myocardial mask overlay

(a)

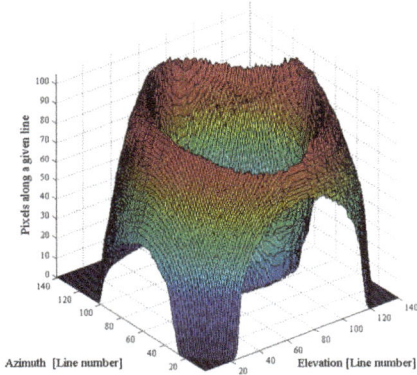

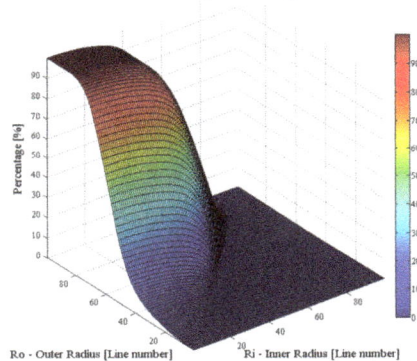

(b)  (c)

**Figure 3.** Example of the workflow to determine the anatomical relevant space: (**a**) Myocardial segmentation shown as a mask overlaid in three orthogonal slices through the middle of the ultrasound volume (from left to right: azimuth, elevation and depth); (**b**) Computed coverage function expressing the percentage of pixels along each image line that belong to the myocardium; (**c**) Percentage of myocardial coverage as a function of the inner radius (*Ri*) and the outer radius (*Ro*) of the ring-shaped template.

For all datasets tested, the mean opening angle (Φ) for the conical scan was 19.7° ± 8.5° while the mean "thickness" of the cone (*d*Φ) was 19° ± 3.4°. Therefore, a reduction of 48.9% in the number of scan lines required to cover the FOV at full-line density was achieved, approximately doubling the frame rate. Given this estimated "mean" conical FOV, the cross-talk energy level of all the MLT setups was calculated. Figure 4 shows an example of the two-way beam profiles in a C-plane used to calculate the inter-beam interference of a 10MLT system, where a beam was transmitted in one direction and received in each of the MLT directions. The cross-talk energy levels of the tested systems are presented in Figure 5. As can be noted, the 8MLT and 9MLT systems reached the pre-defined threshold of −30 dB, while the 10MLT was only half a dB above. The 12MLT system on the other hand was about 4 dB above the cut-off.

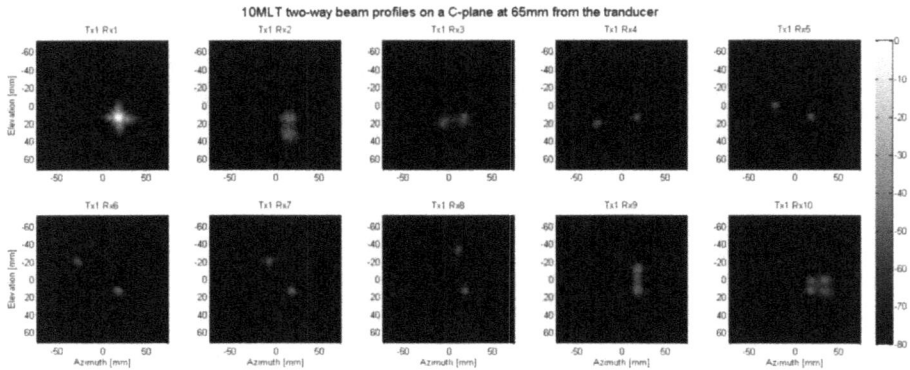

**Figure 4.** Two-way beam profiles of a 10MLT system on a C-plane at 65 mm, when transmitting at one direction (i.e., $\theta_i$) and receiving in all other MLT directions (i.e., $\theta_1,..\theta_i,..\theta_j$). Tukey apodization ($\alpha = 0.5$) was used in transmit and receive. The first subplot can be seen as the single line transmit signal (i.e., the ideal case), and the remaining ones as noise (i.e., cross-talk). This information was used to evaluate Equation (1) which quantifies the cross-talk energy level of the system.

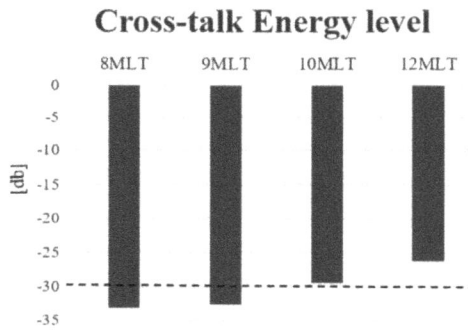

**Figure 5.** Cross-talk energy level of the different tested setups for the estimated 'mean' conical FOV. The tested systems were: 8MLT, 9MLT, 10MLT and 12MLT, all with transmit/receive Tukey apodization ($\alpha = 0.5$). The cross-talk cut-off was defined as $-30$ dB for an acceptable B-mode image quality.

## 5. Discussion

In this study, a methodology to find the anatomical relevant space for functional cardiac imaging has been described. This method was used to compute the parameters for a conical scan and to find an optimal MLT scan sequence to further speed up the acquisition process. In particular, an average opening angle of 19.7° and a mean thickness of 19° for the conical scan were found, resulting in 48.9% less transmit events per volume and, therefore, a gain in frame rate of about 2. In comparison, simply reducing the pyramidal FOV of current 3D systems to the epicardium in end-diastole would only result in a reduction of 25.6% in the number of transmit events or a gain in frame rate of about 1.3. This is expected, given the significant size of the inner cone, which would be subtracted in the presented approach.

Clearly, these values depend on the threshold $T$ defining the amount of myocardial coverage that should be retained. The lower the coverage, the higher the gain in frame rate but the more myocardium that remains outside the FOV and can therefore not be analyzed for function. Thus, this threshold should be chosen with caution and in consideration of the specific application. Although $T$ was considered fixed in this study, it could also be chosen dynamic. Hereby, one would accommodate for the fact that the myocardial wall thickness changes during the cardiac cycle, which implies that the FOV is expected to be dynamic as well. If $T$ is constant, this implies that the line density would have to be variable during the cardiac cycle in order to keep the acquisition time of a single frame constant. Alternatively, line density is kept fixed implying that the $T$ would dynamically need adjustments to keep the total scan time for a given anatomic volume the same. In this study, the $T$ value was arbitrarily set fixed to 85% but could be modified as required. Nevertheless, this arbitrary threshold allowed to provide initial insight into the characteristics of the conical FOV to be encountered in a clinical setting. It should be mentioned that the ring-shape template matching used to determine the opening angles of the conical volume might not be ideal given the left ventricular geometry. As such, an elliptic template might be more adequate but as this also complicates the MLT setup, this extension remains to be investigated in future work and was considered outside the scope of this initial feasibility study on anatomical volumetric imaging.

In this proof-of-concept study, the segmentation was performed at the end-diastole phase only to obtain the largest epicardial mask possible as an upper limit of the outer cone to limit the FOV. However, this could imply that the inner cone, of the field of view was underestimated as the endocardial border moves inwards during systole. Ideally, the segmentation is therefore envisioned to be dynamic throughout the cardiac cycle in order to adjust the FOV dynamically as well.

Given the characteristics of the typical conical volumes to be encountered in a clinical setting, the 8MLT, 9MLT and 10MLT systems seemed to be able to generate high frame rate images while preserving an appropriate image quality as these implementations reached our predefined threshold of $-30$ dB cross-talk. Even though this threshold was at first chosen from retrospectively analyzing our MLT findings in 2D [22], our recent 3D in vitro experiments corroborated that this threshold is indeed a reliable proxy for good image quality [13]. In fact, very recently Ramalli et al. [23] demonstrated both in vitro and in vivo that a 12MLT system can generate 2D images of clinically diagnostic value while this system had inter-beam cross-talk levels well above our pre-defined $-30$ dB threshold [11]. As such, our $-30$ dB threshold was likely conservative implying that also the 12MLT system tested in the context of anatomical imaging remains an attractive option that should be evaluated experimentally.

Overall, the combination of MLT and the definition of an anatomic FOV would thus imply a gain in frame rate up to 24 (considering 12MLT an option). As it has previously been proposed that at least 4MLA (two in azimuth and two in elevation) can be reconstructed per focused beam for a typical 2D cardiac matrix array transducer [14], this implies that the proposed sequences could effectively gain a factor up to 96 in frame rate at full line density. Similarly, if an MLT-4MLA sequence is implemented an actual frame rate gain factor of 64, 72 and 80 could be obtained when using 8 MLT, 9 MLT and 10MLT beams respectively. Finally, in order to further improve the temporal resolution of the data set, line density and/or interpolating the received scanline data could be considered. As such, if required for the desired cardiac application, frame rates in the order of hundreds or even thousands of Hertz could be achieved if a sparse anatomical scan sequence is implemented. For example, according to the current guideline for cardiac chamber quantification [1], for the assessment of regional left ventricular (LV) function (e.g., wall motion or regional strain), the ventricle is divided into segments. One of the segmentation models commonly used divides the apex, the basal and midventricular levels, in six segments each. Resulting in an 18-segment model. As such, as little as six beams would be required to scan the left ventricle with a clinically relevant spatial sampling (i.e., one sample per LV segment). Thus, using MLT beamforming, a single transmit event could evaluate all LV segments defined this way at a volume rate of 5 kHz (i.e., equal to the Pulse Repetition Frequency (PRF) typically used for cardiac imaging). However, six beams around the left ventricular myocardium may be too few to correctly resolve cardiac kinematics. As a consequence, a trade-off between the frame rate and the spatial sampling of the volume has to be made. For instance, if the 12MLT-4MLA sequence is applied to scan the FOV, 48 image lines could be reconstructed per transmit event. If five transmit events are then performed, 240 samples around the circumference could be acquired at a frame rate of 1 kHz. Depending on the application, the transmit events could be performed radially or circumferentially, while properly staggering the scanning sequence should ensure that the inter-beam distance remains constant. Although such sequence would result in inhomogeneous spatial sampling (i.e., transmit beams are further apart than MLA beams), the MLA beams could be used to assess local kinematics (e.g., 3D motion) while the MLT beams allow for sampling different segments of the LV simultaneously.

Only MLT systems with 8 to 12 parallel beams were evaluated in this study, as this was the range of number of beams that gave a significant increase in frame rate while having a moderate cross-talk energy level. It can be expected that systems with less than eight parallel beams will have a better image quality; however, due to their low impact on the frame rate they were not considered in this study. Furthermore, given that the inter-beam interference of the 12MLT system was above the predefined threshold, it can be expected that adding more MLT beams will only result in worse image quality.

In order to translate the proposed scan sequence to the clinical setting, several approaches could be envisioned. For example, one could consider a fixed conical volume based on a statistical analysis of what is likely to be encountered in a clinical setting, and define an optimal MLT sequence for this volume as described in the present study. However, given the large variability of ventricular size and shape, a fairly large FOV would be required implying that only a relatively moderate increase in frame rate would be achieved.

A better alternative would be to ask the user to first identify the FOV in the pyramidal volume acquired using conventional imaging. This FOV could then be used to set up the fast MLT imaging sequence. Nevertheless, this manually delineated FOV adapted to the patients' anatomy would still remain relatively large to ensure that the myocardium remains inside during the full cardiac cycle. Evidently, as there is a time lag between finding the FOV and defining the imaging scan sequence, the latter would not have to be done in real-time (but sufficiently fast) and different MLT setups could prospectively be linked to characteristic dimensions of the conical volume in a look-up table (LUT).

Finally, a third—and preferred—alternative would be to use a dynamic FOV definition based on a fully automatic 3D ultrasound myocardial segmentation. These technologies are already available, e.g., [24,25]. This would lead to an optimal, narrow FOV, and thus an optimal frame rate gain. Given the simplicity of defining the proposed coverage function (whose calculation can be parallelized in a straightforward manner), and given the possibility of real-time template matching [26], the anatomical FOV could be dynamically defined. Then, the best-fitting parallelized scan sequence can be looked up in a LUT which could—similar to the LUT used in the manual FOV definition—prospectively be defined for a set of predefined dynamics FOVs. Of note in this context is that MLT beamforming has recently been demonstrated to be possible in real time [23] and does not require off-line image reconstruction.

Thus, the ideal workflow for anatomical cardiac imaging could be envisioned as:

i     A pyramidal volume is acquired at a conventional frame rate (i.e., ~30 Hz).
ii    Using any existing automatic real-time segmentation framework (e.g., BEAS [24] or real-time contour tracking library, RCTL [25]), the left ventricular myocardium is detected.
iii   The coverage function and the ring-shaped template matching are applied to define an anatomical conically shaped FOV.
iv    Based on (iii), a fast scanning sequence is automatically selected using a LUT giving the best combination of transmit and receive parallelization (i.e., MLT-MLA) to scan the detected region-of-interest.
v     The anatomical relevant space is scanned at high spatiotemporal resolution for subsequent motion analysis.

The proposed approach offers approximately twice the frame rate of the pyramidal 3D MLT approach previously presented by our team [17], despite the fact that it uses less parallelization (i.e., 12MLT-4MLA vs. 16MLT-4MLA). This improvement is obviously due to the reduced FOV to the anatomical relevant space. However, the effective gain in frame rate can only be quantified after choosing the scan sequence based on the desired application as pointed out above.

Please note that the presented methodology was defined for apical view acquisitions only. This is because most of the 3D cardiac kinematic analysis, such as motion and deformation imaging, require a volume scanned from the apex. Although, parasternal view acquisitions are mostly used to asses aortic and pulmonary valves, they are also used for circumferential strain evaluation which is not possible to be appraised with the current approach.

Due to mainly hardware limitations, in vitro and in vivo experiments of a MLT implementation could not be studied and therefore the capability of MLT-MLA systems to precisely record the motion of the heart could not be corroborated. On the one hand, transmitting into several directions simultaneously requires the application of non-identical electric excitation pulses (EEPs) to all elements in the phased array transducer. These pulses are defined as the sum of the EEPs that would be applied on the individual elements when emitting the transmit beams separately. As this sum will result in the superposition of the EEPs, to date, MLT implementations were done on systems having arbitrary waveform generators (AWG). This is particularly relevant for 3D imaging as the superposition of EEPs can become significant (e.g., up to 16 for a 16MLT imaging sequence). However, as most of the available scanners are equipped with a tri-state pulse generator, the implementation of an MLT sequence on such scanner is not straightforward. One way of overcoming this limitation is by clipping the superposed MLT sequences to the maximum of the available pulser. It has been found in preliminary tests that the clipped EEPs are capable of well-resembling the ideal MLT sequence, despite the drop of the transmit energy which might reduce the capability of the acoustic wave to propagate in depth and might decrease the acoustic pressure. On the other hand, matrix array transducers (essential to 3D imaging) have between 2000 and 4000 individual elements which makes wiring costly and impractical. Several methods have been proposed to limit cabling between the US probe and the back-end of the console. Among these methods is subaperture beamforming (SAP) which is most commonly implemented in clinical scanners. SAP divides the transducer aperture in subsets of elements that are beamformed inside the probe and the resulting signal is transferred through a single wire. In this way, the number of wires required to connect the transducer and the back-end of the console are reduced to the number of "sub-apertures". Therefore, beamforming is typically done in two phases where part of the beamforming is done in the transducer itself while the pre-beam formed signals are combined in the front-end in order to generate the individual image lines (i.e., two-stage beamforming [27]). Recently, it has been shown that severe imaging artifacts are introduced while using this technique for DWI [28]. This is due to delay errors which cause grating lobes and reduced penetration. Thus, due to the complexity of MLT sequences, an adaptation for clinical transducers using SAP must be implemented, which represents a challenge that needs to be investigated. Moreover, current state-of-the-art commercial systems are fully equipped to beamform MLA beams in real time. Therefore, the translation of the proposed method into commercial systems remains challenging, since this would not only require changing imaging methodology but also redesigning hardware.

## 6. Conclusions

Anatomical scanning in combination with MLT-MLA beamforming techniques can increase frame rate significantly while keeping information of the relevant structures for functional myocardial imaging. When limiting the FOV to a cone, a frame rate gain factor of about 2 could be achieved. Furthermore, when combined with parallel transmit/receive beamforming and transmit/receive apodization (Tukey: $\alpha = 0.5$), frame rate could be further increased with a 64-fold, 72-fold or 80-fold factor in the case of an 8MLT-4MLA, 9MLT-4MLA or 10MLT-4MLA configuration, respectively, while showing acceptable cross-talk levels. Although the tested 12MLT-4MLA configuration, with the same apodization, did not reach the predefined cross-talk limit, its 96-fold frame rate gain might outweigh its loss in image quality. It therefore remains a possible implementation to be investigated in future in vivo and/or in vitro experiments.

**Acknowledgments:** The research leading to these results has received funding from the European Research Council under the European Union's Seventh Framework programme (FP7/2007–2013)/ERC Grant Agreement number 281748.

**Author Contributions:** Alejandra Ortega under the supervision and guidance of Jan D'hooge, developed the presented methodology, performed the experiments and wrote the paper. Joao Pedrosa and Brecht Heyde contributed with the segmentation masks used in the experiments, and Ling Tong contributed with her knowledge about parallel beamforming implemented in the methodology. All coauthors contributed with reading and improving the manuscript.

**Conflicts of Interest:** The authors declare no conflict of interest.

## References

1. Lang, R.M.; Badano, L.P.; Mor-Avi, V.; Afilalo, J.; Armstrong, A.; Ernande, L.; Flachskampf, F.A.; Foster, E.; Goldstein, S.A.; Kuznetsova, T.; et al. Recommendations for cardiac chamber quantification by echocardiography in adults: An update from the American society of echocardiography and the European association of cardiovascular imaging. *Eur. Heart J. Cardiovasc. Imaging* **2015**, *16*, 233–271. [CrossRef] [PubMed]
2. Jasaityte, R.; Heyde, B.; D'hooge, J. Current state of three-dimensional myocardial strain estimation using echocardiography. *J. Am. Soc. Echocardiogr.* **2013**, *26*, 15–28. [CrossRef] [PubMed]
3. D'hooge, J.; Konofagou, E.; Jamal, F.; Heimdal, A.; Barrios, L.; Bijnens, B.; Thoen, J.; Van de Werf, F.; Sutherland, G.R.; Suetens, P. Two-dimensional ultrasonic strain rate measurement of the human heart in vivo. *Ultrason. Ferroelectr. Freq. Control. IEEE Trans.* **2002**, *49*, 281–286. [CrossRef]
4. Kanai, H.; Koiwa, Y. Myocardial rapid velocity distribution. *Ultrasound Med. Biol.* **2001**, *27*, 481–498. [CrossRef]
5. Pernot, M.; Fujikura, K.; Fung-Kee-Fung, S.D.; Konofagou, E.E. ECG-gated, mechanical and electromechanical wave imaging of cardiovascular tissues in vivo. *Ultrasound Med. Biol.* **2007**, *33*, 1075–1085. [CrossRef] [PubMed]
6. Shattuck, D.P.; Weinshenker, M.D.; Smith, S.W.; von Ramm, O.T. Explososcan: A parallel processing technique for high speed ultrasound imaging with linear phased arrays. *J. Acoust. Soc. Am.* **1984**, *75*, 1273–1282. [CrossRef] [PubMed]
7. Sarvazyan, A.P.; Rudenko, O.V.; Swanson, S.D.; Fowlkes, J.B.; Emelianov, S.Y. Shear Wave Elasticity Imaging: A new ultrasonic technology of medical diagnostic. *Ultrasound Med. Biol.* **1998**, *24*, 1419–1435. [CrossRef]
8. Provost, J.; Papadacci, C.; Arango, J.E.; Imbault, M.; Fink, M.; Gennisson, J.-L.; Tanter, M.; Pernot, M. 3D ultrafast ultrasound imaging in vivo. *Phys. Med. Biol.* **2014**, *59*, L1–L13. [CrossRef] [PubMed]
9. Walker, W.F.; Trahey, G.E. A fundamental limit on the performance of correlation based phase correction and flow estimation techniques. In Proceedings of the IEEE Ultrasonics Symposium, Montréal, QC, Canada, 23–27 August 2004.
10. Prieur, F.; Dénarié, B.; Austeng, A.; Torp, H. Multi-Line Transmission in Medical Imaging Using the Second-Harmonic Signal. *IEEE Trans. Ultrason. Ferroelectr. Freq. Control* **2013**, *60*, 2682–2692. [CrossRef] [PubMed]
11. Tong, L.; Gao, H.; D'hooge, J. Multi-transmit beamforming for fast cardiac imaging-a simulation study. *IEEE Trans. Ultrason. Ferroelectr. Freq. Control.* **2013**, *60*, 1719–1731. [CrossRef] [PubMed]
12. Demi, L.; Verweij, M.D.; Van Dongen, K.W. Parallel transmit beamforming using orthogonal frequency division multiplexing applied to harmonic imaging—A feasibility study. *IEEE Trans. Ultrason. Ferroelectr. Freq. Control* **2012**, *59*, 2439–2447. [CrossRef] [PubMed]
13. Demi, L.; Ramalli, A.; Giannini, G.; Mischi, M. In Vitro and in Vivo tissue harmonic images obtained with parallel transmit beamforming by means of orthogonal frequency division multiplexing. *IEEE Ultrason. Symp. Proc.* **2015**, *62*, 230–235. [CrossRef] [PubMed]
14. Hergum, T.; Bjastad, T.; Kristoffersen, K.; Torp, H. Parallel Beamforming Using Synthetic Transmit Beams. *IEEE Trans. Ultrason. Ferroelectr. Freq. Control* **2007**, *54*, 271–280. [CrossRef] [PubMed]
15. Santos, P.; Tong, L.; Ortega, A.; Løvstakken, L.; Samset, E.; D'hooge, J. Acoustic Output of Multi-Line Transmit Beamforming for Fast Cardiac Imaging. *IEEE Trans. Ultrason. Ferroelectr. Freq. Control* **2015**, *62*, 1320–1330. [CrossRef] [PubMed]

16. Tong, L.; Ortega, A.; Gao, H.; D'hooge, J. Fast three-dimensional ultrasound cardiac imaging using multi-transmit beamforming: A simulation study. *IEEE Ultrason. Symp. Proc.* **2013**, *60*, 1456–1459.

17. Ortega, A.; Provost, J.; Tong, L.; Santos, P.; Heyde, B.; Pernot, M.; D'hooge, J. A comparison of the performance of different multi-line transmit setups for fast volumetric cardiac ultrasound. *IEEE Trans. Ultrason. Ferroelectr. Freq. Control* **2016**, *63*, 1. [CrossRef] [PubMed]

18. Barbosa, D.; Dietenbeck, T.; Schaerer, J.; D'hooge, J.; Friboulet, D.; Bernard, O. B-spline explicit active surfaces: An efficient framework for real-time 3-D region-based segmentation. *IEEE Trans. Image Process.* **2012**, *21*, 241–251. [CrossRef] [PubMed]

19. Pedrosa, J.; Barbosa, D.; Heyde, B.; Schnell, F.; Rösner, A.; Claus, P.; D'hooge, J. Left Ventricular Myocardial Segmentation in 3-D Ultrasound recordings: Effect of Different Endocardial and Epicardial Coupling Strategies. *IEEE Trans. Ultrason. Ferroelectr. Freq. Control* **2017**, *64*, 525–536. [CrossRef] [PubMed]

20. Bruyneel, T.; Ortega, A.; Tong, L.; D'hooge, J. A GPU-based implementation of the spatial impulse response method for fast calculation of linear sound fields and pulse-echo responses of array transducers. Proceedings of IEEE Ultrasonic Symposium, Prague, Czech Republic, 21–25 July 2013.

21. Rademakers, F.; Engvall, J.; Edvardsen, T.; Monaghan, M.; Sicari, R.; Nagel, E.; Zamorano, J.; Ukkonen, H.; Ebbers, T.; Di Bello, V.; et al. Determining optimal noninvasive parameters for the prediction of left ventricular remodeling in chronic ischemic patients. *J. Scand. Cardiovasc.* **2013**, *47*, 329–334. [CrossRef] [PubMed]

22. Tong, L.; Ramalli, A.; Jasaityte, R.; Tortoli, P.; D'hooge, J. Multi-transmit beamforming for fast cardiac imaging—Experimental validation and in vivo application. *IEEE Trans. Med. Imag.* **2014**, *33*, 1205–1219. [CrossRef] [PubMed]

23. Ramalli, A.; Dallai, A.; Boni, E.; Bassi, L.; Meacci, V.; Giovannetti, M.; Tong, L.; D'hooge, J.; Tortoli, P. Multi transmit beams for fast cardiac imaging towards clinical routine. In Proceedings of the IEEE International Ultrasonics Symposium (IUS), Tours, France, 18–21 September 2016.

24. Barbosa, D.; Dietenbeck, T.; Heyde, B.; Houle, H.; Friboulet, D.; D'hooge, J.; Bernard, O. Fast and fully automatic 3D echocardiographic segmentation using B-spline explicit active surfaces: Feasibility study and validation in a clinical setting. *Ultrasound Med Biol.* **2013**, *39*, 89–101. [CrossRef] [PubMed]

25. Orderud, F.; Rabben, S.I. Real-time 3D segmentation of the left ventricle using deformable subdivision surfaces. In Proceedings of the IEEE Computer Society Conference on Computer Vision and Pattern Recognition (CVPR), Anchorage, AK, USA, 23–28 June 2008.

26. Jurie, F.; Dhome, M. Real Time Robust Template Matching. In Proceedings of the 13th British Machine Vision Conference, University of Cardiff, Cardiff, Wales, 2–5 September 2002; pp. 1–10.

27. Savord, B.J. Beamforming Methods and Apparatus for Three-Dimensional Ultrasound Imaging Using Two-Dimensional Transducer Array. U.S. Patent 6013032A, 11 January 2000.

28. Santos, P.; Haugen, G.; Løvstakken, L.; Samset, E.; D'hooge, J. Diverging Wave Volumetric Imaging Using Sub-Aperture Beamforming. *IEEE Trans. Med. Imag.* **2016**, *63*, 2114–2124.

*applied*
*sciences*

MDPI

*Article*

# Multi-Plane Ultrafast Compound 3D Strain Imaging: Experimental Validation in a Carotid Bifurcation Phantom

Stein Fekkes *, Anne E. C. M. Saris, Jan Menssen, Maartje M. Nillesen, Hendrik H. G. Hansen and Chris L. de Korte

Medical Ultrasound Imaging Center, Department of Radiology and Nuclear Medicine, Radboud University Medical Center, Nijmegen 6500 HB, The Netherlands; anne.saris@radboudumc.nl (A.E.C.M.S.); jan.menssen@radboudumc.nl (J.M.); m.nillesen@gmail.com (M.M.N.); rik.hansen@radboudumc.nl (H.H.G.H.); chris.dekorte@radboudumc.nl (C.L.d.K.)
* Correspondence: stein.fekkes@radboudumc.nl; Tel.: +31-24-366-8967

Received: 28 February 2018; Accepted: 16 April 2018; Published: 20 April 2018

**Featured Application: Noninvasive patient-friendly detection of vulnerable plaques in 3D segments of superficial arteries in cardiovascular risk populations.**

**Abstract:** Strain imaging of the carotid artery (CA) has demonstrated to be a technique capable of identifying plaque composition. This study assesses the performance of volumetric strain imaging derived from multi-plane acquisitions with a single transducer, with and without displacement compounding. These methods were compared to a reference method using two orthogonally placed transducers. A polyvinyl alcohol phantom was created resembling a stenotic CA bifurcation. A realistic pulsatile flow was imposed on the phantom, resulting in fluid pressures inducing 10% strains. Two orthogonally aligned linear array transducers were connected to two Verasonics systems and fixed in a translation stage. For 120 equally spaced elevational positions, ultrasound series were acquired for a complete cardiac cycle and synchronized using a trigger. Each series consisted of ultrafast plane-wave acquisitions at 3 alternating angles. Inter-frame displacements were estimated using a 3D cross-correlation-based tracking algorithm. Horizontal displacements were acquired using the single probe lateral displacement estimate, the single probe compounded by axial displacement estimates obtained at angles of 19.47 and −19.47 degrees, and the dual probe registered axial displacement estimate. After 3D tracking, least squares strain estimations were performed to compare compressive and tensile principal strains in 3D for all methods. The compounding technique clearly outperformed the zero-degree method for the complete cardiac cycle and resulted in more accurate 3D strain estimates.

**Keywords:** polyvinyl alcohol phantom; bifurcation; dual probe acquisitions; plane wave ultrasound; 3D strain estimation

## 1. Introduction

Ultrasound (US) strain imaging [1] is used for a wide variety of biomedical applications. One of them is functional assessment of the carotid vessel wall. Strain imaging holds the ability to quantify the mechanical deformation of the arterial wall induced by the pulsating blood and provides insight into the composition and morphology of the vessel wall, which is unavailable from ultrasonic imaging alone [2–5]. Therefore, strain imaging could reveal ongoing, frequently asymptomatic, pathological processes like plaque development. This is very important, because eventually plaque could develop into either stable (predominantly thick fibrous tissue) or vulnerable plaque (large thrombogenic lipd core sealed with a thin cap). The latter type is prone to rupture [6,7]. Since rupture of these plaques has been shown to be one of the main initiators of acute transient ischemic attack or stroke [8], it implies that strain imaging might enable up-front risk stratification of these potentially lethal events. With that it will also provide opportunities for early intervention and hopefully aid in the prevention of these events.

Many arterial strain imaging techniques have already been developed and reported, like intravascular techniques [9], which use an US transducer on the tip of a catheter to image from within the lumen, and non-invasive techniques, which image from outside the body [10–15]. With these techniques, it has been shown that arterial strain distribution correlates with histological composition of plaques in pigs [16] and in humans [2,17]. For intravascular imaging and non-invasive imaging in longitudinal planes, the US beam is aligned with the principal direction of deformation of the artery (radial for a concentric homogeneous artery), which implies a direct and accurate estimation of the strains in the principal direction using algorithms that only estimate the (axial) displacement component along the beam. For non-invasive strain estimation in transverse planes, accurate estimation of the full 2D displacement vector requires more advanced methods [18–21].

Until now, all in vivo applied techniques have been primarily based on cross-sectional images of the artery and provide a strain estimation in 2D only. However, as illustrated before, 2D estimation can be problematic in the case of complex motion patterns as a result of longitudinal artery wall displacement [22,23]. It has been demonstrated that strains are generally smaller in the longitudinal direction than in the radial direction, although the displacement magnitude in the longitudinal direction of the arterial wall equals the magnitude of the radial displacements [24]. Reduced longitudinal artery wall motion was even correlated with cardiovascular outcome [25], which makes an estimate of the full 3D tissue motion in horizontal, vertical, and longitudinal direction even more relevant.

Full 3D tissue motion estimation overcomes the following limitations of 2D. First, due to longitudinal wall motion, 2D transverse US acquisitions will suffer from out of plane motion reducing the strain estimation accuracy. A second limitation of 2D scanning is that it is more difficult to find the optimal cross-section needed for appropriate assessment of carotid disease, i.e., the most vulnerable spot might be missed. Also, inter- and intra-operator variability in skill and experience could hamper a reliable judgement about the best response to the therapy of the plaques [26].

Three-dimensional B-mode-based US of the carotid artery [27,28] has been described before and has been shown to improve the visualization of the intima-media thickness. It complements Intima Media Thhickness (IMT), which is a typical 2D US-based biomarker [29] with a measurement of the Total Plaque Volume (TPV) [30] and the Vessel Wall Volume (VWV) [31]. However, the implementation of 3D ultrasound in these studies did not allow measurement of 3D tissue motion or strain. The main aim of the present study is to develop a 3D ultrasound technique that enables 3D vascular strain estimation of the carotid artery. As aforementioned, adding this functionality could possibly lead to a significant improvement of the prediction of plaque vulnerability.

A few studies have investigated 3D strain imaging for vascular applications. In vitro studies have shown a 3D representation of the 2D strain tensor derived from multi-slice 2D displacement estimations [32,33]. In this approach, the dominant tissue motion in radial direction was aligned with the US beam direction due to the rotational acquisition protocol using the longitudinal axis for rotation. However, the rotation applied in this approach cannot be performed noninvasively, and therefore translation towards clinical practice is not realistic. Furthermore, in reality, the vessel wall tends to deform in three orthogonal directions due to the arterial pressure pulse, and the vessel wall dynamics are affected by the heterogeneous composition and complex geometry in the presence of a stenosis or a plaque [24]. Volumetric 2D strain tensor estimation based on 3D motion estimation showed the capability of handling out of plane motion, i.e., the direction perpendicular to the transversal image plane [23]. However, measurement of the complete 3D strain tensor for the entire vessel wall of the carotid artery in 3D is desirable in the pursuit of better understanding the underlying true deformation encompassing the horizontal, vertical, and longitudinal directions. This requires 3D tracking of the 3D tissue deformation and the derivation of the full 3D strain tensor.

To our knowledge, only one study reporting 3D strain tensors for arteries using intravascular B-mode US was presented in cylindrical coordinates [34], resulting in radial, circumferential, and longitudinal strain distributions. However, predefining the coordinate space to be cylindrical assumes an equal cylindrical geometry of the vessel wall to relate to the dominant strain direction. In the case of a geometry such as the bifurcation, this relation does not hold. Principal strains based on the principles of continuum mechanics have the intrinsic characteristic of coordinate system independence and are therefore more suitable to report the full 3D strain distribution at the bifurcation. 2D principal strains were reported for vascular applications based on 2D strain tensor in previous studies [35,36]. In this paper we implemented the acquisition method of our previous simulation study in an experimental setup and extended the strain estimation algorithm to full 3D. To evaluate the performance of this implementation experimentally, we constructed a phantom of the same patient-specific geometry as used in the simulation study of PolyVinyl Alcohol (PVA) solutions, subjected it to a pulsating physiological flow, and evaluated the accuracy of our 3D strain estimation. The main benefit of the simulation study was the availability of a ground truth to compare the performance of different strategies. However, this study did not incorporate all ultrasound artifacts that are present when using real US acquisitions, like reverberations, shadowing, or inhomogeneity of the propagation speed of sound. To overcome these limitations and approximate in vivo situations more closely, the Finite Element Model (FEM) geometry of the simulation study was used to construct a PVA-based phantom using 3D printed molds. This yielded a realistic phantom of the bifurcation, delivering realistic mechanical behavior when subjected to a fluid pressure pulse. However, due to the re-production variability and stability of PVA-cryogel-based phantoms in terms of elastic modulus [37], no real ground truth can be derived from it. Therefore, a multi-perspective dual probe setup was used to derive optimal strain estimates that operate as the ground truth of this work.

Although it has been previously reported that 2D strain estimation based on displacement compounding outperforms zero-degree based 2D strain estimation [38], to our knowledge, the comparison has never been made for a complex geometry using a realistic phantom.

## 2. Materials and Methods

### 2.1. Experimental Setup

A patient-specific vessel wall geometry of a Carotid Artery (CA), including the common carotid, the internal and external carotid, and the bifurcation [23], was used to build an ultrasound-compatible (i.e., matching the acoustical properties of soft tissue) phantom made of PVA-cryogel (Mw 85.0–124.0 99+% hydrolyzed, Sigma-Aldrich, Zwijndrecht, The Netherlands). Two molds were created using CAD (Autodesk Inventor 2015, Student version, San Rafael, CA, USA) to convert the CA-geometry (Figure 1a) and the surrounding body into 3D printable casts. Subsequently, these molds were filled with a pre-heated ~85° homogeneous mixture of 54% by weight (wt. %) distilled water, 36 wt. % anti-freezing agent (ethylene glycol, Sigma-Aldrich, Zwijndrecht, The Netherlands), and 10 wt. % PVA including either 2.0 wt. % or 0.5 wt. % Silica gel particles (Merck Kieselgel 60, 0.063–0.100 mm, Boom B.V., Meppel, The Netherlands) for the CA-geometry and the surrounding tissue, respectively. After injection and degassing for 1 h, the solutions in the molds were subjected to 4 cycles of freezing ($-25\ ^{\circ}$C) and thawing (21 $^{\circ}$C) of 16 and 8 h, respectively. Prior to the last cycle, the surrounding tissue was casted around the pre-fixated carotid artery bifurcation, resulting in the final composed phantom (Figure 1b). This yielded a higher elastic modulus for the CA-geometry compared to the surrounding tissue. The Young's modulus of the CA embedded in surrounding tissue was experimentally estimated to be 225 kPa, which is within range of values corresponding to intima tissue from atherosclerotic plaques [39].

**(a)**

**(b)**

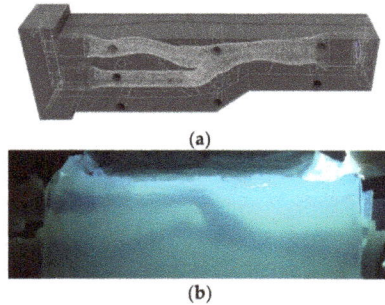

**Figure 1.** (**a**) 3D CAD-based mold of a realistic patient-specific carotid artery and (**b**) Polyvinyl alcohol phantom casted in a surrounding body.

The phantom was connected to an inlet flow tube at the common carotid artery side and to two tubes (internal and external carotid artery) at the outlet side. The connections were supported by a mounting bracket also allowing pressure measurements (TruWave pressure transducer, Edwards Lifesciences Corp., Irvine, CA, USA), see Figure 2b. Fluid was circulated in a closed loop circuit by a gear pump driven by in-house software written in LabView (National Instruments, Austin, TX, USA). The CA was pressurized using a realistic pulsatile flow waveform at 60 bpm with a mean flow of 0.62 L min$^{-1}$ and a peak flow of 1.39 L/min. At both outlet tubes, adjustable flow resistors were set such that the dynamic pressure, measured at the inlet of the CA, was 150 over 100 mmHg, mimicking the pressures encountered in a stenotic carotid artery. With these pressure profiles, strains of up to 10% over the pressure cycle were induced. As can be observed in Figure 2a, the surrounding body contained two orthogonal surfaces for ultrasound imaging, from which the averaged depth to the bifurcation was ~20 mm. Two L12-5 ATL linear array probes were positioned in a probe holder, which allowed fine-tuning to ensure orthogonal fixation and coinciding transversal fields of view. Both probes were closely hovering above the phantom imaging surfaces, while ultrasound gel established the acoustic coupling and simultaneously prevented physical deformation of the phantom due to direct phantom-probe contact. The probe holder was connected to a translation stage enabling elevational movement for 120 equally spaced (0.1 mm) positions while maintaining a constant in plane orientation of the probes. Each transducer was connected to a Verasonics ultrasound system (a Vantage 256 and a V1, Verasonics, Kirkland, WA, USA).

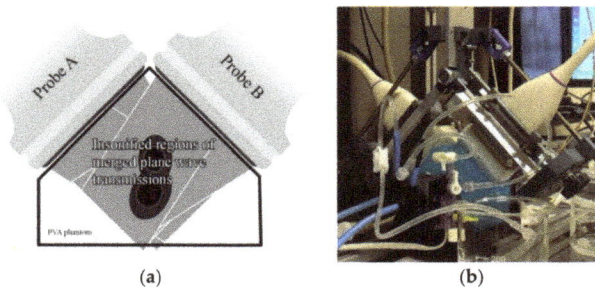

**(a)**

**(b)**

**Figure 2.** (**a**) A schematic drawing of the image acquisition setup in which two probes were fixated orthogonally. The position of the probes was set such that the carotid bifurcation was at a realistic depth and all steered and non-steered insonified regions by the plane wave transmissions encompassed the carotid artery. The phantom was connected to a pressurized realistic dynamic flow circuit; (**b**) The transducer fixation clamp was connected to a translation stage enabling elevational automated movement in a step size of 0.1 mm.

## 2.2. Ultrasound Data Acquisition

Both Verasonics systems and the translation stage were triggered by a common trigger generated by the Labview program at the start of the flow waveform (i.e., the start of the systolic phase). This enabled the time synchronization of the ultrasound data acquired for one cardiac cycle.

At each elevational position ultrasound element data series were acquired at a repetition frequency of 50 Hz for 1 s, i.e., one pressure cycle. Figure 3 depicts the temporal acquisition scheme showing subsequent acquisition series. Each series consists of plane wave acquisitions at alternating angles [23] of $-19.47°$, $19.47°$, and $0°$ for probe A and $0°$ and $-19.47°$ and $19.47°$ for probe B. The order in which these acquisitions took place was chosen such that the time between combined acquisitions in post processing was minimized.

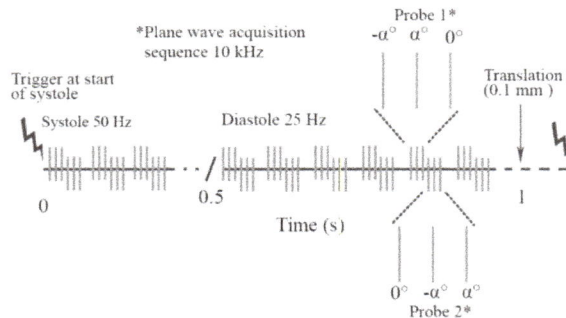

**Figure 3.** Schematic of the temporal acquisition protocol applied using an interleaved dual probe setup.

The steered transmissions were acquired by applying a linear time delay in transmission. The Pulse Repetition Frequency (PRF) of these series was set to 10 kHz, and the tissue was assumed to be in a quasi-static state of deformation, i.e., intra-acquisition series displacements were assumed to be zero, which seems reasonable given the high PRF and the relatively slow motion of the arterial wall. After completing all acquisitions series for the full cardiac cycle, the probes were translated to the next elevational position and the whole sequence was repeated. In this way, a complete volume over the full cardiac cycle was obtained. Finally, frame-skipping was applied at the diastolic phase, yielding an effective PRF of 25 Hz to increase strain estimation signal to noise ratio [40] and to reduce computation time for this phase of the pressure cycle.

## 2.3. Post Processing RF Element Data

Plane wave imaging allows for the conversion of unfocused element data into beamformed RF data at any given location within the region insonified. In this way, not only the axial and lateral grid ratio can be tailored to optimize the displacement estimation accuracy [41], but also the spacing of the beamforming grid can be adjusted to facilitate a fast combination of the steered acquisitions used in the process of displacement compounding as described before in [23]. This beam-forming approach was also used in this study for a steering angle $\alpha$ equal to $19.47°$. An additional orthogonally steered grid ($90°$) was added to enable a match with the $0°$ grid of the other probe. For all four beamforming grid orientations, a single displacement grid was defined. Figure 4 shows the grid layout for both probes, indicating the displacement grid defined by the coinciding points of all grid definitions. The spacing of the beamforming grids and the displacement grid (coinciding points) is listed in Table 1. Element data were beamformed using Delay And Sum (DAS) beamforming assuming a speed of sound of 1540 m/s, a dynamic focus F = 0.875, and Hamming apodization window in receive. Please note that the resolution is specified in terms of dx, dy, and dz in Table 1, in which x is oriented in the lateral direction, y is oriented in the axial direction, and z is oriented in the elevational direction.

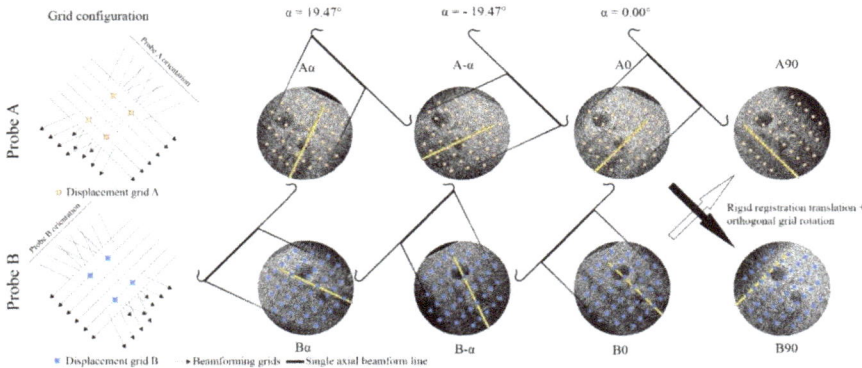

**Figure 4.** The first column depicts the grid configurations for both probes in which the overlay of all steered and non-steered grids results in a displacement grid consisting of coinciding sample-locations. Columns 2, 3, and 4 show the B-mode representations of the actual transmissions performed subsequently overlaid with a displacement grid and a single line indicating the direction of the reconstructed ultrasound beam in receive. The schematic drawing of the aperture indicates the +45° and −45° angles of insonification relative to the phantom for probe A and B, respectively. The last column shows the RF data registration in which the axial direction of the probes serves as a native lateral ultrasound transmission of the opposite probe.

**Table 1.** Resolution of interlocked steered and non-steered grids.

| | Beamforming Grid (μm) | | Displacement Grid (μm) |
|---|---|---|---|
| | Zero-Degree | −α, +α, Orthogonal | |
| dx | 64.2 | 30.2 | 64.2 |
| dy | 10.1 | 10.7 | 9.1 |
| dz | 100 | 100 | 100 |

B-mode representations of the beamformed RF-data of the external and internal carotid artery are shown in Figure 4 for both probes and for steered ($\alpha = \pm 19.47°$) and non-steered ($\alpha = 0.0°$) transmissions. The orthogonal positions of both probes relative to each other allows one to utilize the axial zero-degree estimated displacements of probe A as lateral displacement estimates for probe B and vice versa as depicted in the last column of Figure 4. In this way, highly accurate estimates of the lateral displacments can be obtained, because native RF-phase information can be used in the estimation process, which is normally absent when using a single probe. Finally, the 2D beamformed RF data were stacked according to their elevational position to create a volume matrix of RF data over time. Please regard video S1 (Multi-perspective_3D_Bmode.mp4) for a dynamic B-mode presentation of the carotid artery phantom for both probes.

### 2.4. 3D Interframe Displacement Estimation

Displacement estimations were performed for the RF data beamformed on each of the four grids for each elevational position and time point for both probes. The resulting displacement estimates were labeled using an A or a B and a subscript. The A and B refer to the probe being used. The subscript refers to the acquisition angle and beamforming grid being used. Thus, $A_{-\alpha}$, refers to the displacements estimated using probe A and a beamforming grid and acquisition at steering angle $-\alpha$.

Displacement estimation was performed using a 2-step, normalized Cross-Correlation (CC) based method using 3D kernels defined in Table 2. In the first step, envelope (demodulated RF) data were used to find a global displacement estimate at sample accuracy (i.e., full samples). Subsequently, subsample displacements were determined by 3D spline interpolation of the auto and cross-correlation function based on RF [42]. Median filtering was applied for all three directions using 3D kernels (see Table 2). For the elevational direction, additional regularization was applied by allowing only sub-plane displacement estimations.

**Table 2.** 2-step cross-correlation-based displacement and strain estimation settings.

| Axis Direction | 3D Interframe Displacement Estimation | | | | 3D Strain Estimation | | |
| | Window Size ± Search Range (Beamforming Grid Samples) | | Regularization (Displacement Grid Samples) | | Least Square Window Sizes (Ax, Lat, Ele) | | |
| | Sample Iteration | Sub-Sample Iteration | Median Filtering (Kernel Sample Size Ax, Lat, Ele) | | | | |
| | | | Inter–Frame | Tracked | Ax | Lat | Ele |
| Axial | $81 \pm 15$ | $33 \pm 5$ | 7, 11, 3 | 7, 11, 3 | 17, 7, 3 | 5, 25,3 | 5, 7, 15 |
| Lateral | $13 \pm 4$ | $7 \pm 4$ | 7, 11, 3 | 7, 11, 3 | 17, 7, 3 | 5, 25, 3 | 5, 7, 15 |
| Elevational | $3 \pm 0$ | $3 \pm 2$ | 28, 20, 5 | 7, 11, 3 | 33, 11, 7 | 9, 49, 7 | 9, 11, 61 |

Since probe A and B are positioned orthogonally, and the axial and lateral displacement directions of probe A are equal to the lateral and axial displacement directions of probe B, respectively, and vice versa. The orientation of Probe A is chosen to be the primary orientation throughout the rest of this manuscript. The axial and lateral direction of probe A equals the vertical and horizontal direction of the accumulated displacement estimates as depicted in Figure 5.

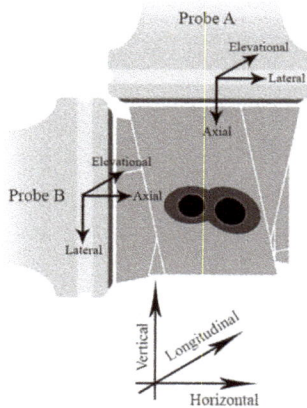

**Figure 5.** The axis definition and nomenclature for the probe related directions: axial, lateral, and elevational. Vertical, horizontal, and longitudinal directions are related to the phantom orientation.

The displacement estimates from both probes for different directions are combined in three different ways used to derive strain estimates. These methods are described below and named: single-probe zero-degree, single-probe compounding, and Dual-probe reference.

2.4.1. Single-Probe Zero-Degree

In the single-probe zero-degree method, the axial ($d_0^{ax}$), lateral ($d_0^{lat}$), and elevational ($d_0^{el}$) inter-frame displacements estimates ($d$) based on either $A_0$ or $B_0$ were used as basis for strain estimation. Please note that the superscript denotes the displacement estimation direction relative to the corresponding probe.

## 2.4.2. Single-Probe Compounding

For the single-probe compounding method, the axial displacement ($d_0^{ax}$), together with the lateral inter-frame displacement estimated by displacement compounding, served as basis for strain estimation. In displacement compounding, both angular axial displacement components ($d_\alpha^{ax}$, $d_{-\alpha}^{ax}$) obtained using the angled transmissions for each probe ($A_{-\alpha}$, $A_\alpha$ for probe A and $B_{-\alpha}$, $B_\alpha$ for probe B) were used to derive the lateral displacement $d_0^{comp\_lat}$:

$$d_0^{Comp\_lat} = \frac{d_\alpha^{ax} - d_{-\alpha}^{ax}}{2 \sin \alpha} \tag{1}$$

Combining the two axial displacements into a single lateral displacement has shown to be beneficial, since a lateral displacement estimate based on the phase information of two angled acquisitions yields a more accurate estimate than the lateral displacement estimation originating from zero-degree acquisitions [18]. To determine the elevational displacement component for the single-probe compounding approach, the displacements $d_0^{ele}$, $d_\alpha^{ele}$, and $d_{-\alpha}^{ele}$ were combined to a single estimate by means of the median displacement. Please note that these displacements are perpendicular to the elevational direction.

## 2.4.3. Dual-Probe Reference

The axial ($d_0^{ax}$) displacement estimate used for this method equals the displacement estimate used for the single-probe zero-degree and single-probe compounding method. For the lateral displacement component ($d_0^{lat}$), the axials $A_0$ & $B_0$ were registered such that they could be used as RF data originating from $A_{90}$ & $B_{90}$, (see Figure 4 last column) preserving phase information in the lateral direction. This method combines 0° axial displacement estimates ($d_0^{ax}$ & $d_{90}^{ax}$) from both *A* & *B* probes, respectively. Since this method uses solely zero-degree axial displacement results, it is expected to provide the most accurate displacement estimates in the transverse plane, and therefore it will serve as reference in this work. For the elevational displacement, $d_0^{el}$ was used.

## 2.5. Regularization Tracking and Segmentation

To accumulate the filtered inter-frame 3D displacements over the complete pressure cycle, displacement tracking was performed, with the end-diastolic frame as reference frame. Hereto axial, lateral, and elevational displacement estimates were accumulated spatially over time using 3D linear interpolation. Additional 3D median filtering was performed on the accumulated displacements, with kernel sizes listed in Table 2.

Finally, a Region of Interest (ROI) at the end diastolic frame was defined by manually contouring the lumen and wall using a composition of all 6 registered B-mode images for all 120 slices in elevational direction. Tracking and strain estimation was only performed for this ROI, representing the vessel wall and its direct surrounding to minimize the computational load.

## 2.6. Strain Estimation

For all three methods, strain estimation was performed using 3D least-squares strain estimation [43] in the axial, lateral, and elevational direction. The applied kernel sizes are listed in Table 2 and were empirically optimized for the reference method, after which they were kept constant for the other two methods. For the calculation of the 3D principal components, an eigenvalue decomposition was performed of the 3 × 3 strain tensor consisting of the normal strains and the shear strain components. At last, the 3 principal components were arranged from maximum to minimum values with corresponding eigenvectors. The maximum and minimum principal strains represent the maximum tensile strain component and the minimum compressive strain component, respectively.

## 2.7. Evaluation

In summary, Figure 6 shows a flowchart indicating the processing steps described in the previous sections from raw ultrasound acquisition, up to principal strain estimation for the zero-degree method, the compounding method, and the reference method.

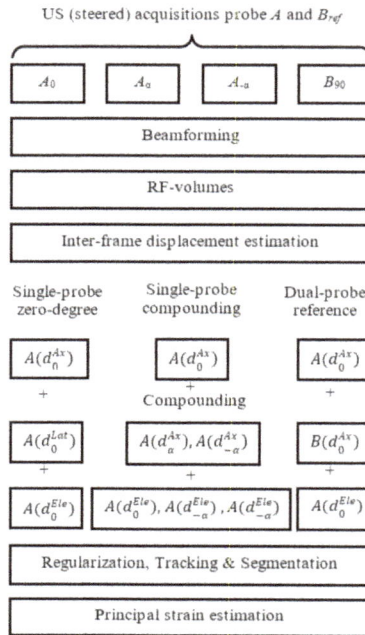

**Figure 6.** Flowchart indicating the in-line processes performed from the acquisitions to strain estimation to compare the single-probe compounding versus the single-probe zero-degree performance in terms of minimal (compressive) and maximum (tensile) 3D principal strains. Please note that this flowchart describes the situation with respect to probe A. A similar flowchart with respect to probe B was also applied and can be obtained by interchanging A and B.

For performance evaluation, the vertical, horizontal, and longitudinal strain results for both probes were quantitatively and qualitatively reported. The final performance evaluation was done by Root Mean Squared Error (RMSE) analysis of the minimum (compressive) and maximum (tensile) principal strains over time.

## 3. Results

Figure 7 shows the vertical and horizontal strain results at peak systolic phase ($t = 0.3$ s) for six distinct elevational slices for the three different methods. The reference results based on the axial strain as obtained with probe A and the axial strain obtained with probe B are presented in the left column. The compounding results and zero-degree single probe results are shown in the center and right column, respectively.

As indicated by the axial probe direction arrows and described in the methods section, the reference results originate from displacement estimations using solely information along the ultrasound beam direction for both probes. Clear resemblance is shown between the reference strains and the compounded strains, although the strains obtained by the compounding method appear noisier in some regions. This in contrast to the zero-degree lateral component, which depicts a highly non-uniform strain distribution with many strain clipping regions delivering unreliable lateral strain distributions.

Figure 8 depicts the elevational strain distributions at peak systolic pressure, for the reference, compounding, and zero-degree method for both probes. All strain distributions show compressive and tensile strains along the longitudinal direction of the vessel wall at the same longitudinal positions. Strains of up to 10% are observed at the external carotid artery for both probe orientations. Intra-probe strain distributions indicate a high similarity between the different methods. Inter-probe comparison, however, reveals a rotation in the strain distribution orientation at the inlet side (communis) and the outlet side (external CA).

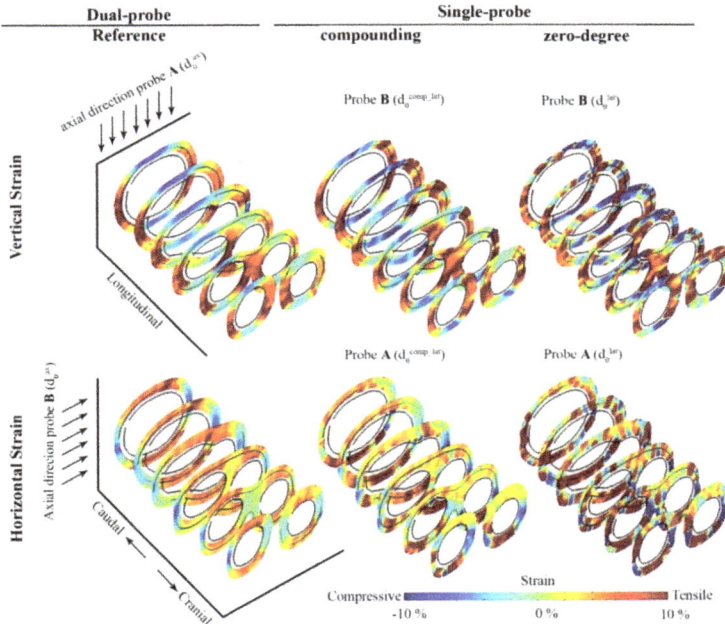

**Figure 7.** Vertical and horizontal strain estimates of the bifurcation in 3D at maximum systolic pressure. Please note that for visualization purposes, only 6 distinctive transverse slices are depicted at a 3 times larger elevational scale.

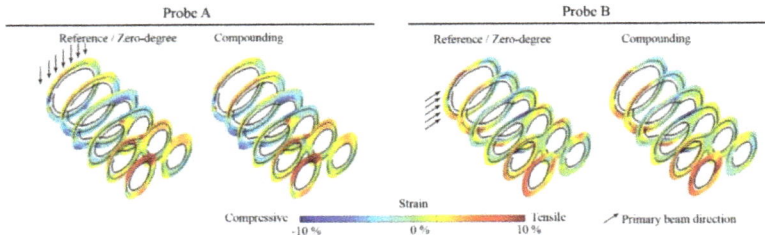

**Figure 8.** Longitudinal strain estimates originating from both probes at peak systolic pressure.

Figure 9 shows the strain distribution over time for one cardiac cycle for the vertical, horizontal, and longitudinal direction for the reference, compounding and zero degree methods for probe A and B. For the vertical strains of probe A and the horizontal strain of probe B, all lines almost perfectly coincide. The 5th and 95th percentile horizontal strain of the zero-degree method show extreme strain values of <−15% and >20% at peak systolic pressure compared to the reference (~−5%, ~8%) strain, respectively. The compounding method outperforms the zero-degree method, which shows a high resemblance to the reference strain distribution for the complete cardiac cycle. For the longitudinal strain distribution, all three methods perform almost equally. Similar trends can be observed for probe B with respect to the reference of probe A.

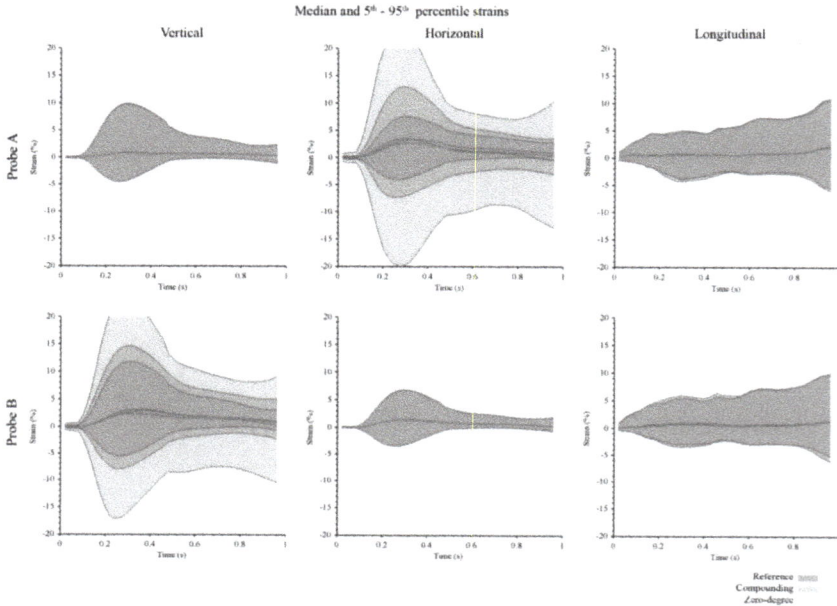

**Figure 9.** The vertical, horizontal, and longitudinal strain range (median and 5th–95th percentile) for the complete cardiac cycle for the reference, compounding, and zero-degree method for probe A and B.

Table 3 lists the absolute median strain and the 5th and 95th percentile strain values for both probes to indicate the strain range at two different time points in the cardiac cycle. For the vertical component of probe A and the horizontal component of probe B, the percentile strains shows almost no difference at maximum systolic pressure for all methods. Please note that these strains originate from axial (Figure 5) displacement estimations for both probes. Also, small residual median strains were measured for these corresponding orientations.

**Table 3.** Vertical, horizontal, and longitudinal 5th–95th percentile strain values for both probes and all methods.

| | Probe | Strain (%) at Max. Systolic Pressure $t = 0.30$ s | | | Residual Strain (%) $t = 0.96$ s | | |
| | | Reference | Compounding | Zero-Degree | Reference | Compounding | Zero-Degree |
|---|---|---|---|---|---|---|---|
| Vertical | A | −4.2–9.7 | −4.3–9.8 | −4.9–9.7 | −1.1–2.3 | −1.1–2.3 | −1.2–2.3 |
| | B | −5.3–11.8 | −7.6–14.6 | −16.0–23.0 | −0.7–3.2 | −2.2–5.1 | −10.5–9.1 |
| Horizontal | A | −3.9–7.5 | −7.0–13.0 | −19.7–23.6 | −0.5–2.8 | −3.1–3.5 | −13.1–0.4 |
| | B | −3.2–6.8 | −3.2–6.7 | −3.2–6.7 | −0.6–1.8 | −0.6–1.9 | −0.6–1.8 |
| Longitudinal | A | −3.8–4.7 | −4.3–5.0 | −3.9–4.7 | −5.6–10.7 | −6.0–11.0 | −5.6–10.7 |
| | B | −3.4–5.5 | −2.9–6.0 | −3.4–5.5 | −6.1–9.9 | −5.2–9.8 | −6.1–9.9 |

For the vertical strain of probe B and horizontal strain of probe A (i.e., lateral direction for each probe, percentile strain values are formatted in bold in Table 3), an increased strain range was found for the compounding technique. However, this range was much more increased for the zero-degree technique. Both strain graphs (Figure 9) show a divergence at $t > 0.8$ s of the strain range. This divergence is observed also for the longitudinal strain range for the complete pressure cycle leading to almost equal residual strain ranges in between ~−6% and ~11% for all methods and both probes.

Eigenvector decomposition of a 3 × 3 strain tensor provides 3 independent eigenvalues and eigenvectors that are perpendicular to each other. Figure 10 shows the minimum (bottom row) and maximum (top row) principal components, representing the most dominant compressive and tensile deformations as estimated by probe A at peak systolic pressure. For a dynamic development of these components for the complete cardiac cycle, please see video S2 (3D_Principal_MinMax_Strains.mp4). For visualization purposes, principal strains (color) and the corresponding eigenvectors (white arrows) are plotted for three distinct longitudinal positions. The top slice is situated in the common carotid artery. The middle slice is at the bifurcation, and the bottom slice shows the internal and external carotid arteries.

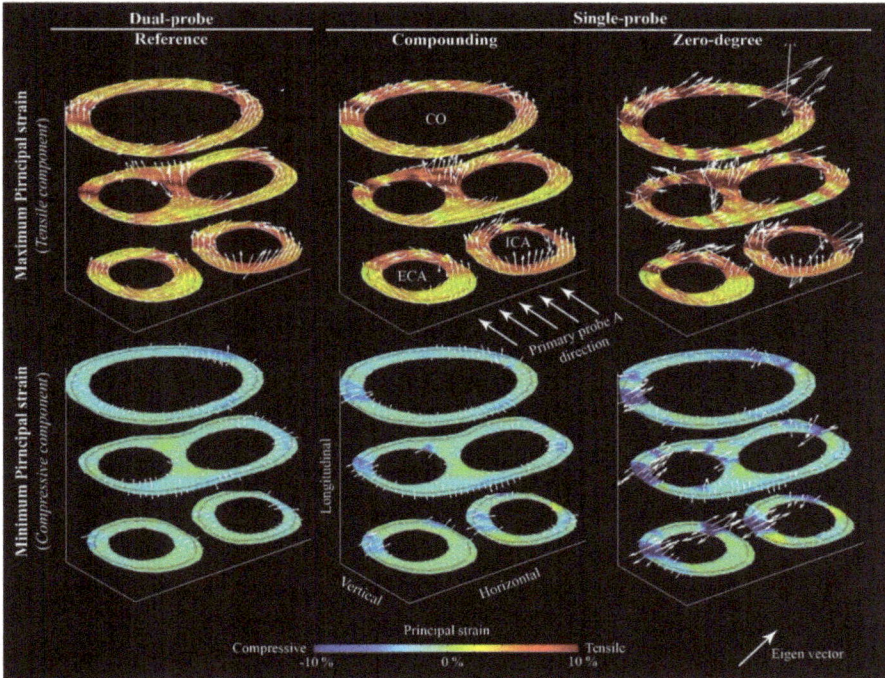

**Figure 10.** Maximum (tensile) and minimum (compressive) principal strain distributions for probe A for the reference, compounding, and zero-degree method in the bifurcation carotid artery region. Principal strains are visualized at the level of the COmmon carotid artery (CO), bifurcation, and distal from the bifurcation, in which the Internal Carotid Artery (ICA) and the External Carotid Artery (ECA) are present. Please note that the lengths of the eigenvectors were scaled to their corresponding eigenvalue.

In general, the tensile strain distribution is clearly oriented in the circumferential direction of the vessel wall with exception of a region in the bifurcation slice and one in the internal carotid artery slice, in which the orientation of the eigenvectors indicates a strong longitudinal preference.

The compressive strain distribution is mostly oriented in the radial direction and is less dominant than the tensile strain distribution. High tensile strains are observed consistently for all methods at the flow divider of the bifurcation and at ten and two o'clock of the common carotid artery.

Smooth values and eigenvectors can be appreciated for the reference method, indicating high tensile strains at the flow divider. Many outliers of the strain values and a chaotic distribution of eigenvectors can be observed for the zero-degree method, clearly indicating the disability in providing consistent and trustworthy results. Especially in the regions in which the reference principal components point in the horizontal direction, erroneous values manifest for the zero-degree method. For the compounding method, the errors in these regions are far less destructive. Although the compounding technique is not perfect, it does yield strain results with less noise compared with the zero-degree method for the complete cardiac cycle, which shows erroneous results primarily in the regions in which the direction of the eigenvectors coincides with the lateral direction for each probe, which for probe A corresponds to the horizontal direction depicted in Figure 10. For the compounding method, the errors in these regions are far less destructive.

Figure 11 shows the RMSE of the minimal (left) and maximum (right) principal strain over time for a single cardiac cycle. The 5th–95th percentile strain ranges for both probes were depicted in the background to relate the RMSE to the overall strain development. In general, the RMSE for the compounding methods is smaller for the complete cycle in comparison with the zero-degree method. However, an up rise of the RMSE and the reference strain range occurs at $t > {\sim}0.8$ s, which indicates a progressive strain error in the process of strain accumulation for all methods.

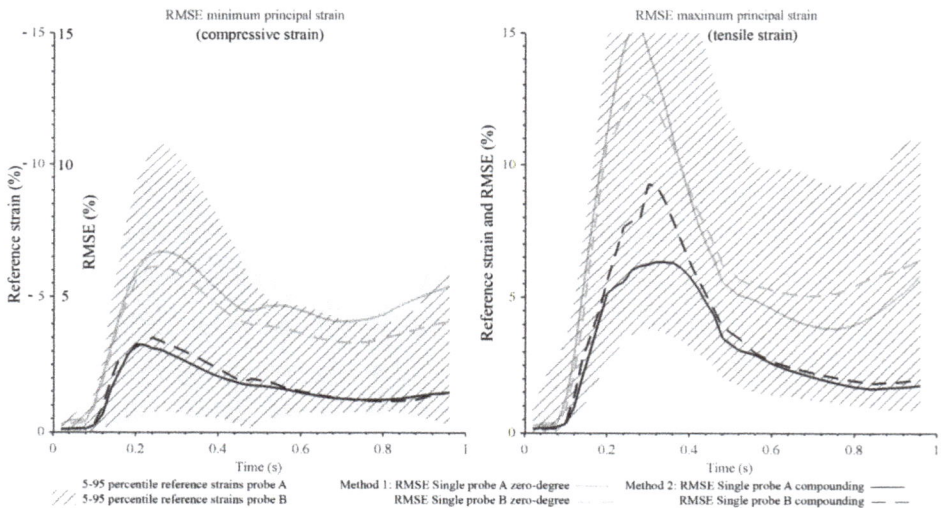

**Figure 11.** The RMSE of the compressive and tensile strain components for the complete cardiac cycle for the zero-degree and compounding method for both probes. Please note the double axis for the RMSE minimal principal strain, in which the RMSE is always positive; the 5th–95th percentiles strain range is negative.

The compressive strain RMSE values for probe A and B using only the zero-degree angle at peak systole are 6.6% and 5.9% versus 2.7% and 3.1% for the compounding method. Also, the tensile strain RMSE shows elevated values for the zero-degree method of 14.3% and 12.6%, in comparison to 9.3% and 6.2% for the compounding method.

Looking at the residual compressive strain RMSE, values of 5.4% and 4.1% versus 1.5% and 1.5% were found for the zero-degree method and the compounding method, respectively. The same trend holds for the residual tensile strain RMSE, which was 5.8% and 6.4% versus 1.8% and 2.0%.

The mean RMSE values of the compressive principal strain over the complete cardiac cycle for the zero-degree method are 4.4% and 3.8%, respectively, versus 1.7% and 1.9% for the compounding method. For the tensile principal strain, 6.9% and 6.8% versus 3.3% and 4.0% were found. Therefore, the compounding method clearly outperforms the zero-degree method in terms of overall tensile and compressive principal strain over time.

## 4. Discussion

In this study, compressive and tensile 3D principal strain estimation in a pulsating carotid artery bifurcation phantom was performed using a compounding and zero-degree displacement estimation method in a dual probe setup. Previous experimental studies have been performed using straight-forward phantoms (i.e., single tube shaped phantoms) and simulations based on 2D and 3D compound strain imaging [19,23] showing a significant improvement compared to a single angle based strain estimation. Multiple 3D phantoms of the carotid artery have been developed lately [44,45]. However, to our knowledge, no other paper has reported 3D principal strain estimation in a complex geometry like the CA bifurcation. The principal finding of this study is that 3D compound strain estimation outperforms single angle zero degree 3D strain estimation, even in a complex geometry, when embedded in surrounding tissue. Compared to other methods, this method is the first showing volumetric 3D strain tensor information for vascular applications retrieved in a non-invasive manner.

The developed dual probe setup approach, which provided a reference for the estimated deformations, is, to our knowledge, also novel and circumvents requiring exact knowledge of mechanical properties and geometry of the tissue and of boundary conditions, which is normally needed when deriving a finite element, model-based ground truth. Therefore, this setup might also be used in future studies to provide reference deformations for in vitro studies using excised tissue in which the mechanical properties are also not known.

To derive the reference strains from the RF data obtained with the dual-probe setup, the RF data acquired with both probes needs to be spatially registered. Since the angle of the probes is set to 90 degrees, only the translation should be resolved, which was performed using a feature-based registration using coherently compounded B-mode images. Although this might seem to be a simple registration, the quality of the registration depends on the choice of registration metric and also on the speed of sound and the assumption of a homogenous speed of sound throughout the phantom. In our case, setting the sound speed to 1540 m/s provided the most optimal registration, which is in line with reported speed of sound values for PVA [37].

In this work, a patient-specific model of the carotid artery at the bifurcation was used to construct a PVA-based phantom. Since the surrounding tissue affects the mechanical behavior of the carotid artery wall [46], the CA was embedded in surrounding PVA instead of water. The addition of surrounding tissue is not only important for obtaining realistic motion and deformation of the carotid artery segment but also resembles realistic in vivo acquisitions, since deterioration of the strain estimates due to side and grating lobes will be present. What was not accounted for in the current phantoms were heterogeneities within the vessel wall and plaque. Especially, the presence of calcifications and induced shadowing is expected to affect strain estimation accuracy.

In the experiment, the cardiac pressure cycle was enforced with 60 bpm. This resulted in a cyclic reproducible deformation of the vessel phantom. In the process of accumulation of inter-frame displacements, inaccuracies and interpolation errors led to an overall progressive displacement error over time, which explains the observed residual strain for all methods at the end of the pressure cycle (Figure 9). The longitudinal progression of the strain distribution (5th–95th percentile strain) over time shows a divergent trend for the full cardiac cycle. Since a ground truth strain distribution in the longitudinal direction is absent, this indicates the lack of a cyclic strain progression for each cardiac cycle. The elevational ultrasound beam width is one order of magnitude larger than the step-size in elevational direction. This results in redundant ultrasound RF data in the elevational direction. In a previous study, we showed the ability to resolve sub-plane elevational displacement estimation [23], but still it remains the direction with the relatively lowest displacement estimation resolution. Furthermore, the geometrical shape of the lumen in combination with this elevational beam overlap between slices might also lead to displacement estimation errors. Inter-slice geometrical lumen changes might dominate the normalized cross-correlation-process, which is supposed to detect inter-frame displacement changes.

Three methods for each probe should yield similar results for the strain estimation in the elevational direction. Figure 8 confirms that, in general, this is the case, since equal strain distributions between the zero-degree and compounding based methods are observed. Nevertheless, local strain differences were found between probe A and B. These were found especially at the inlet side and outlet side, and might be due to different displacement estimation kernel orientations and filter effects.

To calculate the 3D strain components, a 3D Least Square (LSQ) strain estimator was applied. The kernel size determines the tradeoff between outlier suppression and true value detection. Large kernels sizes result in smooth global strain estimates but also in underestimation of the local strain differences. Especially at the edges, underestimation might occur, since the displacement gradient of the true deformation in the vessel wall is the highest at the vessel wall-lumen transition. To compare the performance across the methods, the LSQ kernel size for the single probe compounding and zero degree method was kept equal to that of the reference method. Application of larger kernel sizes for these methods might be favorable to smoothen results without losing essential strain map features. This will be investigated in future studies.

Further research needs to be carried out to fully translate the single probe compounding technique into the clinic. The practical setup needs to be supportive to actually acquire multi-slice US data from the CA in vivo. Due to the curved neck and jaw skin surface, maintaining acoustic contact will be challenging during the complete acquisition range in the elevational direction. A possible solution to this might be to create a thick, pre-shaped layer of ultrasound transparent material, ensuring constant acoustic contact between the transducer and skin surface.

Another challenge for in vivo acquisition is related to the data acquisition. The trigger originating from the pump driver at the start of each systolic phase initiated data acquisition for the complete cardiac cycle. Since the acquisition took 1 second at each position, thus several minutes for the entire acquisition, the phantom motion had to be very reproducible. In vivo, this multi-slice technique might become more difficult to achieve due to heart rate and blood pressure variability and motion artifacts due to, for example, breathing. However, it is already known that heart rate variability in general only affects the diastolic phase and not the systolic phase of the pressure cycle [47], and only one phase of the pressure cycle is required to obtain maximum strain contrast. Thus, for in vivo application we recommend to apply this technique on the systolic phase while the subject is in resting position to minimize blood pressure variations. ECG-triggering will then need to be incorporated to enable synchronization in vivo, and probably the head and neck region should be fixated to minimize motion artefact.

Finally, for in vivo application a reduction in overall acquisition time is favored. The acquisition time can already be halved compared to the acquisition sequence used in this paper, because, as aforementioned, it is not necessary to acquire data for the diastolic phase. With the development of matrix probes with a sufficient footprint, pitch, and number of elements, it might even become possible to acquire volumetric ultrasound and strain data within one cardiac cycle. These probes will also enable displacement compounding in the elevational direction to increase the elevational displacement estimation accuracy. Inherent to the multi-slice acquisition protocol, no beam steering could be applied in the current implementation.

Overall, we expect the availability of the non-invasive volumetric ultrasound data and 3D strain values to open up new possibilities for the anatomical and functional classification of the CA and its pathological, but frequently asymptomatic, processes.

## 5. Conclusions

A triggered, multi-slice, ultrafast ultrasound acquisition protocol enabled the acquisition of spatially and temporally coherent volumetric ultrasound RF data in a pulsating, patient-specific carotid bifurcation phantom embedded in surrounding material at a frame rate of 50 Hz. The compounding of axial displacements acquired at angles of $0°$, $19.5°$, and $-19.5°$ enabled the estimation of more accurate 3D compressive and tensile principal strains over a pressure cycle than could be obtained from axial and lateral displacements estimates obtained from zero degree acquisitions.

**Supplementary Materials:** The following are available online at http://www.mdpi.com/2076-3417/8/4/637/s1, Video S1: Multi-perspective_3D_Bmode.mp4; Video S2: 3D_Principal_MinMax_Strains.mp4

**Acknowledgments:** This research is supported by the Dutch Technology Foundation STW (NKG 12122), which is part of the Netherlands Organization for Scientific Research (NWO), and which is partly funded by the Ministry of Economic Affairs.

**Author Contributions:** S.F., H.H.G.H., A.E.C.M.S., M.M.N., and C.L.d.K. conceived and designed the project and experiments; J.M. and A.E.C.M.S. contributed to the experimental setup; S.F. performed the experiments; S.F. and H.H.G.H. designed the method and analyzed the data; S.F. drafted the manuscript; H.H.G.H. revised and edited the manuscript. All authors reviewed and approved the manuscript.

**Conflicts of Interest:** The authors declare no conflict of interest.

## References

1. Ophir, J.; Céspedes, I.; Ponnekanti, H.; Yazdi, Y.; Li, X. Elastography: A quantitative method for imaging the elasticity of biological tissues. *Ultrason. Imaging* **1991**, *13*, 111–134. [CrossRef] [PubMed]
2. Hansen, H.H.; de Borst, G.J.; Bots, M.L.; Moll, F.L.; Pasterkamp, G.; de Korte, C.L. Validation of noninvasive in vivo compound ultrasound strain imaging using histologic plaque vulnerability features. *Stroke* **2016**, *47*, 2770–2775. [CrossRef] [PubMed]
3. Ohayon, J.; Finet, G.; Le Floc'h, S.; Cloutier, G.; Gharib, A.M.; Heroux, J.; Pettigrew, R.I. Biomechanics of atherosclerotic coronary plaque: Site, stability and in vivo elasticity modeling. *Ann. Biomed. Eng.* **2014**, *42*, 269–279. [CrossRef] [PubMed]
4. Cardinal, M.H.R.; Heusinkveld, M.H.G.; Qin, Z.; Lopata, R.G.P.; Naim, C.; Soulez, G.; Cloutier, G. Carotid artery plaque vulnerability assessment using noninvasive ultrasound elastography: Validation with MRI. *Am. J. Roentgenol.* **2017**, *209*, 142–151. [CrossRef] [PubMed]
5. Huang, C.W.; He, Q.; Huang, M.W.; Huang, L.Y.; Zhao, X.H.; Yuan, C.; Luo, J.W. Non-invasive identification of vulnerable atherosclerotic plaques using texture analysis in ultrasound carotid elastography: An in vivo feasibility study validated by magnetic resonance imaging. *Ultrasound Med. Biol.* **2017**, *43*, 817–830. [CrossRef] [PubMed]
6. Schaar, J.A.; Muller, J.E.; Falk, E.; Virmani, R.; Fuster, V.; Serruys, P.W.; Colombo, A.; Stefanadis, C.; Ambrose, J.A.; Moreno, P.; et al. Terminology for high-risk and vulnerable coronary artery plaques. *Eur. Heart J.* **2004**, *25*, 1077–1082. [CrossRef] [PubMed]
7. Lendon, C.L.; Davies, M.J.; Born, G.V.R.; Richardson, P.D. Atherosclerotic plaque caps are locally weakened when macrophage density is increased. *Atherosclerosis* **1991**, *87*, 87–90. [CrossRef]
8. Hellings, W.E.; Peeters, W.; Moll, F.L.; Piers, S.R.D.; van Setten, J.; Van der Spek, P.J.; de Vries, J.P.P.M.; Seldenrijk, K.A.; De Bruin, P.C.; Vink, A.; et al. Composition of carotid atherosclerotic plaque is associated with cardiovascular outcome a prognostic study. *Circulation* **2010**, *121*, 1941–1950. [CrossRef] [PubMed]
9. De Korte, C.L.; Céspedes, E.I.; van der Steen, A.F.W.; Pasterkamp, G.; Bom, N. Intravascular ultrasound elastography: Assessment and imaging of elastic properties of diseased arteries and vulnerable plaque. *Eur. J. Ultrasound* **1998**, *7*, 219–224. [CrossRef]

10. Ribbers, H.; Lopata, R.G.; Holewijn, S.; Pasterkamp, G.; Blankensteijn, J.D.; de Korte, C.L. Noninvasive two-dimensional strain imaging of arteries: Validation in phantoms and preliminary experience in carotid arteries in vivo. *Ultrasound Med. Biol.* **2007**, *33*, 530–540. [CrossRef] [PubMed]

11. McCormick, M.; Varghese, T.; Wang, X.; Mitchell, C.; Kliewer, M.A.; Dempsey, R.J. Methods for robust in vivo strain estimation in the carotid artery. *Phys. Med. Biol.* **2012**, *57*, 7329–7353. [CrossRef] [PubMed]

12. Schmitt, C.; Soulez, G.; Maurice, R.L.; Giroux, M.F.; Cloutier, G. Noninvasive vascular elastography: Toward a complementary characterization tool of atherosclerosis in carotid arteries. *Ultrasound Med. Biol.* **2007**, *33*, 1841–1858. [CrossRef] [PubMed]

13. Kawasaki, T.; Fukuda, S.; Shimada, K.; Maeda, K.; Yoshida, K.; Sunada, H.; Inanami, H.; Tanaka, H.; Jissho, S.; Taguchi, H.; et al. Direct measurement of wall stiffness for carotid arteries by ultrasound strain imaging. *J. Am. Soc. Echocardiogr.* **2009**, *22*, 1389–1395. [CrossRef] [PubMed]

14. Poree, J.; Chayer, B.; Soulez, G.; Ohayon, J.; Cloutier, G. Noninvasive vascular modulography method for imaging the local elasticity of atherosclerotic plaques: Simulation and in vitro vessel phantom study. *IEEE Trans. Ultrason. Ferroelectr. Freq. Control* **2017**, *64*, 1805–1817. [CrossRef] [PubMed]

15. Khamdaeng, T.; Luo, J.; Vappou, J.; Terdtoon, P.; Konofagou, E.E. Arterial stiffness identification of the human carotid artery using the stress-strain relationship in vivo. *Ultrasonics* **2012**, *52*, 402–411. [CrossRef] [PubMed]

16. De Korte, C.L.; Sierevogel, M.; Mastik, F.; Strijder, C.; Velema, E.; Pasterkamp, G.; van der Steen, A.F.W. Identification of atherosclerotic plaque components with intravascular ultrasound elastography in vivo: A yucatan pig study. *Circulation* **2002**, *105*, 1627–1630. [CrossRef] [PubMed]

17. Kanai, H.; Hasegawa, H.; Ichiki, M.; Tezuka, F.; Koiwa, Y. Elasticity imaging of atheroma with transcutaneous ultrasound preliminary study. *Circulation* **2003**, *107*, 3018–3021. [CrossRef] [PubMed]

18. Hansen, H.H.; Lopata, R.G.; Idzenga, T.; de Korte, C.L. Full 2d displacement vector and strain tensor estimation for superficial tissue using beam-steered ultrasound imaging. *Phys. Med. Biol.* **2010**, *55*, 3201–3218. [CrossRef] [PubMed]

19. Hansen, H.H.; Saris, A.E.; Vaka, N.R.; Nillesen, M.M.; de Korte, C.L. Ultrafast vascular strain compounding using plane wave transmission. *J. Biomech.* **2014**, *47*, 815–823. [CrossRef] [PubMed]

20. Korukonda, S.; Nayak, R.; Carson, N.; Schifitto, G.; Dogra, V.; Doyley, M.M. Noninvasive vascular elastography using plane-wave and sparse-array imaging. *IEEE Trans. Ultrason. Ferroelectr. Freq. Control* **2013**, *60*, 332–342. [CrossRef] [PubMed]

21. Poree, J.; Garcia, D.; Chayer, B.; Ohayon, J.; Cloutier, G. Noninvasive vascular elastography with plane strain incompressibility assumption using ultrafast coherent compound plane wave imaging. *IEEE Trans. Med. Imaging* **2015**, *34*, 2618–2631. [CrossRef] [PubMed]

22. Larsson, M.; Verbrugghe, P.; Smoljkic, M.; Verhoeven, J.; Heyde, B.; Famaey, N.; Herijgers, P.; D'Hooge, J. Strain assessment in the carotid artery wall using ultrasound speckle tracking: Validation in a sheep model. *Phys. Med. Biol.* **2015**, *60*, 1107–1123. [CrossRef] [PubMed]

23. Fekkes, S.; Swillens, A.E.S.; Hansen, H.H.G.; Saris, A.E.C.M.; Nillesen, M.M.; Iannaccone, F.; Segers, P.; de Korte, C.L. 2-d versus 3-d cross-correlation-based radial and circumferential strain estimation using multiplane 2-d ultrafast ultrasound in a 3-d atherosclerotic carotid artery model. *IEEE Trans. Ultrason. Ferroelectr. Freq. Control* **2016**, *63*, 1543–1553. [CrossRef] [PubMed]

24. Cinthio, M.; Ahlgren, A.R.; Bergkvist, J.; Jansson, T.; Persson, H.W.; Lindstrom, K. Longitudinal movements and resulting shear strain of the arterial wall. *Am. J. Physiol. Heart Circ. Physiol.* **2006**, *291*, H394–H402. [CrossRef] [PubMed]

25. Svedlund, S.; Eklund, C.; Robertsson, P.; Lomsky, M.; Gan, L.M. Carotid artery longitudinal displacement predicts 1-year cardiovascular outcome in patients with suspected coronary artery disease. *Arterioscler. Thromb. Vasc. Biol.* **2011**, *31*, 1668–1674. [CrossRef] [PubMed]

26. Landry, A.; Spence, J.D.; Fenster, A. Quantification of carotid plaque volume measurements using 3d ultrasound imaging. *Ultrasound Med. Biol.* **2005**, *31*, 751–762. [CrossRef] [PubMed]

27. Steinke, W.; Hennerici, M. Three-dimensional ultrasound imaging of carotid artery plaques. *J. Cardiovasc. Technol.* **1989**, *8*, 15–22.

28. Fenster, A.; Parraga, G.; Bax, J. Three-dimensional ultrasound scanning. *Interface Focus* **2011**, *1*, 503–519. [CrossRef] [PubMed]

29. Baldassarre, D.; Amato, M.; Bondioli, A.; Sirtori, C.R.; Tremoli, E. Carotid artery intima-media thickness measured by ultrasonography in normal clinical practice correlates well with atherosclerosis risk factors. *Stroke* **2000**, *31*, 2426–2430. [CrossRef] [PubMed]

30. Landry, A.; Spence, J.D.; Fenster, A. Measurement of carotid plaque volume by 3-dimensional ultrasound. *Stroke* **2004**, *35*, 864–869. [CrossRef] [PubMed]

31. Delcker, A.; Diener, H.C. 3d ultrasound measurement of atherosclerotic plaque volume in carotid arteries. *Bildgeb. Imaging* **1994**, *61*, 116–121.

32. Boekhoven, R.W.; Rutten, M.C.M.; van Sambeek, M.R.; van de Vosse, F.N.; Lopata, R.G.P. Towards mechanical characterization of intact endarterectomy samples of carotid arteries during inflation using echo-ct. *J. Biomech.* **2014**, *47*, 805–814. [CrossRef] [PubMed]

33. Boekhoven, R.W.; Rutten, M.C.M.; van Sambeek, M.R.; van de Vosse, F.N.; Lopata, R.G.P. Echo-computed tomography strain imaging of healthy and diseased carotid specimens. *Ultrasound Med. Biol.* **2014**, *40*, 1329–1342. [CrossRef] [PubMed]

34. Liang, Y.; Zhu, H.; Friedman, M.H. Measurement of the 3d arterial wall strain tensor using intravascular b-mode ultrasound images: A feasibility study. *Phys. Med. Biol.* **2010**, *55*, 6377–6394. [CrossRef] [PubMed]

35. Hansen, H.H.; de Borst, G.J.; Bots, M.L.; Moll, F.L.; Pasterkamp, G.; de Korte, C.L. Compound ultrasound strain imaging for noninvasive detection of (fibro)atheromatous plaques: Histopathological validation in human carotid arteries. *JACC Cardiovasc. Imaging* **2016**, *9*, 1466–1467. [CrossRef] [PubMed]

36. Nayak, R.; Huntzicker, S.; Ohayon, J.; Carson, N.; Dogra, V.; Schifitto, G.; Doyley, M.M. Principal strain vascular elastography: Simulation and preliminary clinical evaluation. *Ultrasound Med. Biol.* **2017**, *43*, 682–699. [CrossRef] [PubMed]

37. Fromageau, J.; Gennisson, J.L.; Schmitt, C.; Maurice, R.L.; Mongrain, R.; Cloutier, G. Estimation of polyvinyl alcohol cryogel mechanical properties with four ultrasound elastography methods and comparison with gold standard testings. *IEEE Trans. Ultrason. Ferroelectr. Freq. Control* **2007**, *54*, 498–509. [CrossRef] [PubMed]

38. Hansen, H.H.G.; Lopata, R.G.P.; Idzenga, T.; De Korte, C.L. Fast strain tensor imaging using beam steered plane wave ultrasound transmissions. In Proceedings of the 2010 IEEE International Ultrasonics Symposium (IUS), San Diego, CA, USA, 11–14 October 2010; pp. 1344–1347.

39. Akyildiz, A.C.; Speelman, L.; Gijsen, F.J.H. Mechanical properties of human atherosclerotic intima tissue. *J. Biomech.* **2014**, *47*, 773–783. [CrossRef] [PubMed]

40. Varghese, T.; Ophir, J. A theoretical framework for performance characterization of elastography: The strain filter. *IEEE Trans. Ultrason. Ferroelectr. Freq. Control* **1997**, *44*, 164–172. [CrossRef] [PubMed]

41. Saris, A.E.C.M.; Nillesen, M.M.; Fekkes, S.; Hansen, H.H.G.; de Korte, C.L. Robust blood velocity estimation using point-spread-function-based beamforming and multi-step speckle tracking. In Proceedings of the 2015 IEEE International Ultrasonics Symposium (IUS), Taipei, Taiwan, 21–24 October 2015; pp. 1–4.

42. Kim, S.; Aglyamov, S.R.; Park, S.; O'Donnell, M.; Emelianov, S.Y. An autocorrelation-based method for improvement of sub-pixel displacement estimation in ultrasound strain imaging. *IEEE Trans. Ultrason. Ferroelectr. Freq. Control* **2011**, *58*, 838–843. [PubMed]

43. Kallel, F.; Ophir, J. A least-squares strain estimator for elastography. *Ultrason. Imaging* **1997**, *19*, 195–208. [CrossRef] [PubMed]

44. Chee, A.J.Y.; Ho, C.K.; Yiu, B.Y.S.; Yu, A.C.H. Walled carotid bifurcation phantoms for imaging investigations of vessel wall motion and blood flow dynamics. *IEEE Trans. Ultrason. Ferroelectr. Freq. Control* **2016**, *63*, 1852–1864. [CrossRef] [PubMed]

45. Boekhoven, R.W.; Rutten, M.C.M.; van de Vosse, F.N.; Lopata, R.G.P. Design of a fatty plaque phantom for validation of strain imaging. In Proceedings of the 2014 IEEE International Ultrasonics Symposium, Chicago, IL, USA, 3–6 September 2014; pp. 2619–2622.

46. Liu, Y.; Dang, C.; Garcia, M.; Gregersen, H.; Kassab, G.S. Surrounding tissues affect the passive mechanics of the vessel wall: Theory and experiment. *Am. J. Physiol. Heart C* **2007**, *293*, H3290–H3300. [CrossRef] [PubMed]

47. Berne, R.M.; Levy, M.N. *Physiology*, 3rd ed.; Mosby Year Book: Saint Louis, France, 1993.

*applied*
*sciences*

MDPI

*Article*

# Fast Volumetric Ultrasound B-Mode and Doppler Imaging with a New High-Channels Density Platform for Advanced 4D Cardiac Imaging/Therapy

Lorena Petrusca [1,*], François Varray [2], Rémi Souchon [3], Adeline Bernard [2], Jean-Yves Chapelon [3], Hervé Liebgott [2], William Apoutou N'Djin [3] and Magalie Viallon [1,2]

1 University of Lyon, UJM-Saint-Etienne, INSA, CNRS UMR 5520, INSERM U1206, CREATIS, F-42023 Saint-Etienne, France; magalie.viallon@creatis.insa-lyon.fr
2 University of Lyon, UCBL, INSA, UJM-Saint Etienne, CNRS UMR 5520, INSERM U1206, CREATIS, F-69100 Lyon, France; Francois.Varray@creatis.insa-lyon.fr (F.V.); Adeline.Bernard@creatis.insa-lyon.fr (A.B.); Herve.Liebgott@creatis.insa-lyon.fr (H.L.)
3 LabTAU, INSERM, Centre Léon Bérard, Université Lyon 1, Univ Lyon, F-69003 Lyon, France; remi.souchon@inserm.fr (R.S.); Jean-Yves.Chapelon@inserm.fr (J.-Y.C.); apoutou.ndjin@inserm.fr (W.A.N.)
* Correspondence: Lorena.Petrusca@creatis.insa-lyon.fr; Tel.: +33-472431988

Received: 22 December 2017; Accepted: 25 January 2018; Published: 29 January 2018

**Abstract:** A novel ultrasound (US) high-channels platform is a pre-requisite to open new frontiers in diagnostic and/or therapy by experimental implementation of innovative advanced US techniques. To date, a few systems with more than 1000 transducers permit full and simultaneous control in both transmission and receiving of all single elements of arrays. A powerful US platform for implementing 4-D (real-time 3-D) advanced US strategies, offering full research access, is presented in this paper. It includes a 1024-elements array prototype designed for 4-D cardiac dual-mode US imaging/therapy and 4 synchronized Vantage systems. The physical addressing of each element was properly chosen for allowing various array downsampled combinations while minimizing the number of driving systems. Numerical simulations of US imaging were performed, and corresponding experimental data were acquired to compare full and downsampled array strategies, testing 4-D imaging sequences and reconstruction processes. The results indicate the degree of degradation of image quality when using full array or downsampled combinations, and the contrast ratio and the contrast to noise ratio vary from 7.71 dB to 2.02 dB and from 2.99 dB to −7.31 dB, respectively. Moreover, the feasibility of the 4-D US platform implementation was tested on a blood vessel mimicking phantom for preliminary Doppler applications. The acquired data with fast volumetric imaging with up to 2000 fps allowed assessing the validity of common 3-D power Doppler, opening in this way a large field of applications.

**Keywords:** ultrasound; 4-D; cardiac; fast volumetric imaging; platform; advanced imaging; power doppler

## 1. Introduction

In the last 15 years, open research US machines [1] have been developed for therapy and imaging in order to offer different solutions to the ultrasound (US) community. Although the US imaging modality is considered a versatile, well-established and widely used diagnostic tool in medicine, novel approaches based on modern signal and image processing methods are required in order to improve the image quality and diagnostic accuracy. But the lack of flexibility offered by standard commercial systems, especially the limited access of RF raw data information and limited number of elements, does not always fit the requirements for data access and extensive control over imaging and systems parameters [2] and prevents a complete evaluation of new investigation methods [3]. Moreover, 3-D imaging systems provide a detailed view of tissue structures that make diagnosis easier for physicians. Therefore, volumetric US scanners with open access, that enable real time 3-D visualization of dynamic structures, including the heart, are required by the community.

A novel US open platform with high-channels density with optimal design constraints such as flexibility, precision, hardware efficiency, optimal acquisition architecture and wide access to RF data is a pre-requisite to open new frontiers in diagnostic and/or therapy by experimental implementation of innovative advanced US techniques: dual-mode US imaging/therapies in the heart, new approaches to study the myocardial tissue (structure/characterization), fast sparse array strategies [4,5], multi-line transmit (MLT) [6], adaptive beamforming, and powerful motion correction strategies, etc. [7–10]. This flexibility is not available in commercial equipment designed for clinical use.

Few experimental high research systems in the world [11,12] today permit full control both in transmission and receiving of all single elements simultaneously of arrays with more than 1000 transducers, requiring as many channels as the number of active elements. One of them, the SARUS scanner [12], is designed to allow full exploitation of 2-D ultrasound transducers, but heavy hardware is required, and therefore its portability is a weak point. In vivo 3-D results, including Doppler applications and cardiac fiber orientation, have already been published by using the parallelized Aixplorer systems [13,14]. For this setup, in order to synthetize a total of 1024 receiving channels, each emission must be repeated twice, with the receiving channels being multiplexed to 1 of 2 transducer elements. Therefore, this platform is not completely open and it has limited flexibility.

Other approaches have been investigated to allow addressing a large amount of active elements with a reduced number of channels: multiplexing [15] by using capacitive micromachined ultrasonic transducer (CMUT) technology [16], micro-beamforming [17] and row-column addressing [18]. But in this case, the flexibility of the sequence acquisition is reduced while the elements coupled in sub-arrays are not continuously connected to the US scanner, limiting the modularity of the system. Furthermore, 2-D optimized sparse arrays strategies were also proposed and evaluated, a continuous one-element-to-one-channel connectivity being provided [5].

The quantitative applications like tissue Doppler imaging or strain imaging when using conventional echocardiography used in clinical routine are limited due to the low frame-rate acquisitions. High frame-rate ultrasound imaging (also called ultrafast) represents a modality developed in the last fifteen years due to its great interest and numerous possible applications [9]. Around 150 frames per second are available when using conventional imaging techniques, while a frame-rate of more than 1000 frames per second is available with plane wave imaging [19]. The interest of achieving a high frame-rate in 3-D echography is highly important for clinical diagnosis, especially for moving organs like the heart. By improving the temporal resolution of the acquired images, the different phases of the cardiac cycle can be visualized and quantitative structural and functional information analyzed. Different methods based on focused or unfocused transmit beams were developed for ultrafast echocardiography, including plane waves (PW) [20], diverging waves (DW) [21] or MLT [6,22] approaches. For increased contrast and spatial resolution, the combination of several images is used for these techniques. Therefore, in this case, motion compensation algorithms [23,24] are required to increase the image quality affected by large tissue displacements. Moreover, complex blood flow quantification [25] or artery elasticity assessment [26] could also be analyzed with ultrafast methods. When using 3-D field of view at a high frame-rate, full quantitative Doppler flow analysis can be performed on a large region of interest, leading to much more information and improved functionality for the clinician [27]. In addition, the accuracy of 3-D power Doppler can also be analyzed.

This paper presents a new powerful fully programmable US platform developed with specific performance for implementing 4-D (real-time 3-D) advanced US strategies with custom data processing. It includes 4 US systems with 256 channels each synchronized together and a US probe with a matrix array of $32 \times 32$ elements. The main features of this imaging system include its capability of simultaneous arbitrary waveform generation, full access to the RF data and the possibility of performing 4-D dual-mode US imaging and/or therapy. In addition, the platform is fully modular up to 1024 elements in full or downsampled regular or sparse array, portable and appropriate for diagnostic and/or therapy purposes. The technical implementation and validation of this platform is described here. Moreover, the feasibility of the fast volumetric US B-mode and Power Doppler imaging approach is also described, opening a large field of vascular and cardiac applications.

## 2. Materials and Methods

### 2.1. US Experimental Platform

The new ultrasound platform includes a 1024-element US probe with a specific wiring of the elements and a high-channels density US scanner driven by 4 individual Vantage 256 systems (Verasonics, Kirkland, WA, USA), synchronized together. The development of this US platform was possible thanks to a close collaboration between two laboratories, by sharing together the required materials (the US probe and four US systems) and their competences. The materials were used together for all the experiments presented in this work.

**US probe.** The customized US probe (Vermon, Tours, France) (Figure 1d) for cardiac applications is a $32 \times 35$ matrix array with a 0.3 mm pitch in both $x$ and $y$ directions and a square footprint with sides of about 10 mm. In the $y$ direction, the 9th, 17th and 25th lines are not connected, resulting in a total number of 1024 active elements.

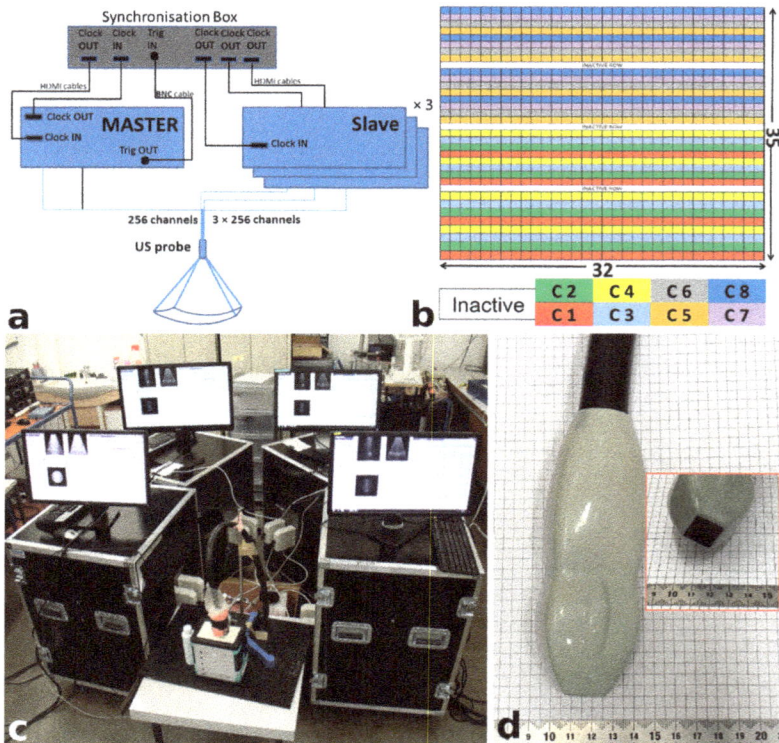

**Figure 1.** (**a**) Block diagram of the entre setup including the Master, the 3 Slaves, the synchronization box and the US probe; (**b**) Diagram of the 1024-element distribution across 8 connectors. Each color corresponds to a connector (C); (**c**) Picture of the 4 synchronized Vantage systems and the 3-D US probe connected; (**d**) the customized 3-D US cardiac probe.

The 1024 elements of the probe were carefully spread out across the 4 systems, with 2 connectors per system, each element being physically connected to the same single channel both in transmission and reception (one-element-to-one-channel design). The physical addressing of each US element was specifically chosen for allowing various array sparsity combinations while minimizing the number of required driving systems. A detailed diagram illustrating the element distribution is shown in Figure 1b). Indeed, each connector of 128 channels controls 4 entire rows (of 32 elements each) distributed across every 4 rows in the matrix. In this way, for a full active array, 4 systems are required, but regular or irregular sparsity of the US array can be activated using a lower number of US systems. All active channels can be used simultaneously in transmission and reception. Moreover, the connectors incorporate impedance matching circuits for power transmission, if required. The central frequency was chosen as 3 MHz with a 70.9% bandwidth of −6 dB.

Therefore, the physical addressing of the US prototype was determined as a compromise between the simplicity of internal connection, flexibility for testing various downsampled configurations and the reduction of the number of systems needed. The interest of several simple sparsity configurations to generate exploiTable 4-D US images was investigated in simulation and then validated experimentally.

**Ultrasound systems.** The Vantage 256 scanners US systems (Verasonics, Kirkland, WA, USA) drive 256 elements each (2 connectors of 128 elements per board in each system) and are a research system entirely programmable (data acquisition and processing platform), designed to allow the development of new advanced US imaging strategies. They have per-channel arbitrary waveform transmit/receive generation capability with full access to RF data from every channel. The systems can be controlled by using Matlab through open user interface, allowing access to RF data of all channels and in particular, to display reconstructed data for ultrasound guidance in real time and offering extreme flexibility and precision. Standard or customized US transducers arrays can be connected to the systems, after declaration of the specifications of the US probe.

**Synchronization box.** A special Multi-System Synchronization Module capable of distributing the Master system clock to up to 8 Vantage systems via common HDMI cables designed and commercialized by Verasonics was used. One of the 4 systems is the "Master", providing the 250 MHz clock, trigger and other signals for accurate temporal synchronization. Therefore, the module triggers in transmission and reception all the connected systems (considered as "Slaves") to co-start the acquisitions, with a less than 10 ns delay. Parallel acquisition can then be performed, each system collecting the data from the 256 elements of the connected sub-probe. This method allows control of up to 1024 elements at once. A block-diagram of the entire setup is illustrated in Figure 1a, while a picture of the experimental setup is presented in Figure 1c.

## 2.2. Numerical Simulations

Numerical simulations (in Field II simulation Program [28,29]) of US imaging were performed first in order to test various configurations and appropriately choose the best ones for the final US probe design. The possible configurations of the US matrix probe, including full and regular downsampled arrays, were tested and the image quality analyzed. The media whose images have been simulated included (i) equidistant point markers, (ii) a cyst or (iii) a numerical phantom with 500,000 scatters having uniform spatial distribution and amplitudes determined from an MR image of the human heart.

## 2.3. Initial Experimental Acquisitions for Validation of the Platform

The 4 Vantage systems were synchronized together for data acquisitions in 6 different array configurations allowed by the US probe. The validation of the setup was performed on a standard phantom (Figure 2) by using PW and DW. The used US acquisition parameters are described in Table 1.

**Table 1.** US acquisition parameters.

| Parameters | Value |
|---|---|
| *Probe parameters* Matrix probe number of elements | 32 × 32 |
| Pitch | 0.3 mm |
| Center frequency | 3 MHz |
| *Imaging parameters* Transmit center frequency | 3 MHz |
| Sampling frequency | 12 MHz |
| Max imaging depth | 60 mm |
| Transmit full aperture Pulse Repetition Frequency (PRF) | 4 × 256 elements 200–2000 Hz |
| Frame-rate | 200–2000 fps |

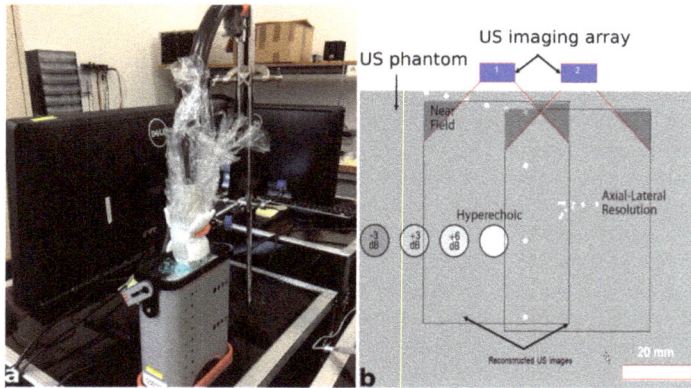

**Figure 2.** (**a**) Validation setup on a standard cyst phantom; (**b**) The diagram of the standard phantom used for experiments. 1 and 2 indicates 2 possible positions of the probe.

Table 2 presents all tested matrix configurations of the US probe for data acquired both with transmissions of plane waves (PW) and diverging waves (DW). The numbers of the connectors (C) are from 1 to 8, while the Verasonics (V) systems are numbered from 1 to 4. For symmetry reasons regarding the center of the probe along the $y$ direction, for the presented examples, the connectors are plugged as follows: C4 and C5 on V1, C3 and C6 on V2, C2 and C7 on V3, while C1 and C8 on V4.

**Table 2.** US probe matrix configurations and Verasonics combinations.

| Array Configuration | Matrix # Columns × #Rows | No. of Active Elements | Connectors (C) Used | No. of Verasonics (V) Systems |
|---|---|---|---|---|
| Full array | 32 × 32 | 1024 | C1:C8 | V1:V4 |
| 1 line ON/1 line OFF | 32 × 16 | 512 | C2,C4,C5,C7 | V1, V3 |
| 2 lines ON/2 lines OFF | 32 × 16 | 512 | C2,C4,C5,C7 | V1, V2 |
| 4 lines ON/4 lines OFF | 32 × 16 | 512 | C1:C8 | V1:V4 |
| Middle ON | 32 × 16 | 512 | C1:C8 | V1:V4 |
| 1 line ON/3 lines OFF | 32 × 8 | 256 | C2,C7 | V2 |

For the validation of the US platform, serial experiments were performed with PW and DW with 9 transmission waves (9 in the $x$ and $y$ direction, resulting in a total of 81 transmissions) by using the matrix configurations described before. An excitation of 30 V was applied in all cases. The US probe was maintained in a fixed position in order to strictly image the same volume of the phantom for a correct comparison of the image quality obtained with the different configurations. Further image quality analysis and the presence of secondary lobes were carried out to compare the results obtained with all configurations. For a qualitative evaluation of the results, the axial and lateral resolution was measured in $(x, z)$ and $(y, z)$ central slices of the phantoms. The full width at half maximum (FWHM) for 3 different wires positioned at 10, 20 and 30 mm was computed and the obtained values analyzed. Moreover, the contrast ratio (CR) and the contrast to noise ratio (CNR) were also analyzed, transformed to decibel (dB) scale. The CNR is a measure of the signal level in the presence of noise, expressing the fact that detectability increases with increasing object contrast and decreasing acoustic noise. An inside region was defined in a $(y, z)$ plane as a circle of radius 6.6 mm edge inside the cyst (8 mm diameter cyst), while an outside region of the same dimensions was chosen above the cyst. The same inside and outside regions were preserved for all the measurements. The CR and CNR were computed according to:

$$CR = 20 \log_{10} \left( \frac{\mu_{in}}{\mu_{out}} \right)$$

$$CNR = 20 \log_{10} \left( \frac{|\mu_{in} - \mu_{out}|}{\sqrt{0.5 \times (\sigma_{in}^2 + \sigma_{out}^2)}} \right)$$

where $\mu_{in}/\mu_{out}$ and $\sigma_{in}/\sigma_{out}$ correspond to the respective mean and standard deviation of the beamformed signals (of the B-mode image module) values inside/outside the cyst.

### 2.4. Experimental Data Acquisitions for Doppler Application

The feasibility of the 4-D platform implementation was tested on a blood vessel mimicking phantom [30] for 3-D Doppler applications. The phantom is made in polyvinyl alcohol (PVA), silica and distilled water (10%, 1% and 89% in weight, respectively), ensuring elasticity (from the PVA) and a good visualization (from the silica) in US images, with 5 cycles of freeze-thaw. The length of the phantom was around 8 cm, while the phantom thickness and internal diameter were 2 mm and 8 mm, respectively. A circulating liquid (water and corn flour) obtained by using a setup with an external pump mimicked the blood flow inside the vessel. The role of the corn flour dispersed in the water increased the blood backscattering in order to quantify the flow. The resulting phantom is close to biological tissues in terms of acoustic and mechanical characteristics [31] and the fluid is in accordance with the standard physical and acoustic properties of blood [32].

The phantom was immersed in static water for acoustic coupling, and the US cardiac probe described in Section 2.1 was placed a few centimeters above. Different set-ups between the vessel mimicking phantom and the US probe were investigated: US probe parallel to the tube phantom ($\theta = 0$) and tilted at an angle of 12°.

The acquisitions for 3-D Power Doppler evaluation were performed using the full active 1024 matrix array of the US probe, with a single PW transmitted. The volumes were acquired at a high frame-rate with 2000 Hz for each configuration.

### 2.5. Data Processing

The RF data were acquired individually for each of the 4 US systems from the active elements to which it was connected to. After the acquisition, the data were collected and merged together, carefully taking into account the inactive line distribution of the US probe in each system. One large matrix containing the information of all the active elements (up to 1024 elements) was obtained and the beamformed RF data were post-processed offline [33]. A dataset for an entire volume was further provided.

For Power Doppler visualisation of the images, 3D slicer software [34] was used, allowing 3-D view rendering, and the animations are shown in supplementary material.

## 3. Results

### 3.1. Preliminary Numerical Simulations

Preliminary numerical simulations allowed estimating the degradation of the images induced by several levels of sparsity (or matrix covering) implemented in 1-D. Figure 3 illustrates 5 different configurations of the active elements, including the full active array and downsampled array with 512 or 256 elements, in the $(y, z)$ plane. The central slice of the simulated volume is illustrated here. As expected, the full array (a) leads to the best visual image quality, while for the configuration with only 256 active elements (only 1 US system required), the image quality is very poor (e) with large secondary lobes and poor contrast. This result is also confirmed by axial resolution measurements in the $(x, z)$ and $(y, z)$ planes, displayed in Table 3. This step allowed us to define the best configuration of the final US probe design. It can be noticed that, when the 512 elements are positioned with an important pitch between the active lines (configuration 4 lines ON/4 lines OFF), the overall slice quality is decreased in the $(y, z)$ plane compared to the two other 512 arrays.

**Table 3.** Axial resolution (the mean value) measured in $(x, z)$ and $(y, z)$ planes for the equidistant scatters when using different configurations of the matrix array.

| Axial Resolution (mm) | | Full Array | 1 Line ON/1 Line OFF | 4 Lines ON/4 Lines OFF | 2 Lines ON/2 Lines OFF | 1 Line ON/3 Lines OFF |
|---|---|---|---|---|---|---|
| $(y, z)$ plane | 40 dB | 1.2 | 2 | 4.6 | 2.6 | 2.2 |
| | −6 dB | 0.7 | 0.8 | 0.9 | 0.6 | 0.6 |
| $(x, z)$ plane | 40 dB | 1.2 | 1.2 | 1.2 | 1.2 | 1.2 |
| | −6 dB | 0.6 | 0.6 | 0.5 | 0.6 | 0.5 |

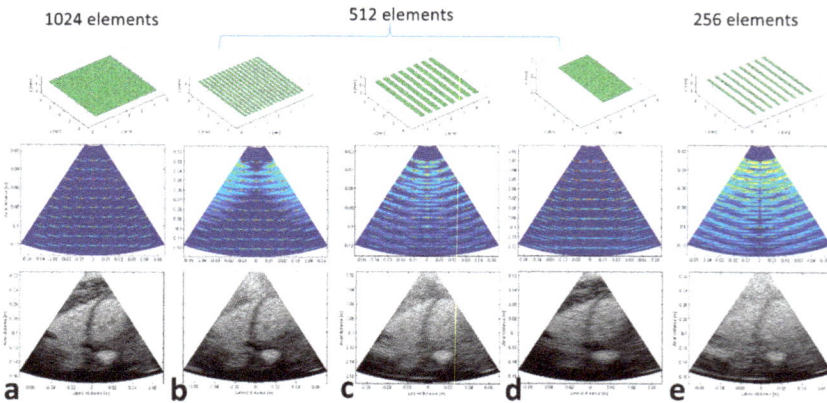

**Figure 3.** Numerical simulations results from Field II on equidistant scatters (2nd row) and based on an MR image of the heart (3rd row) corresponding to different configurations of the US array (1st row) in $(y, z)$ plane.

### 3.2. Experimental Validation of the 4-D US Platform Developped

An example of two different 3-D volumes obtained experimentally with 1024-channels US platform is illustrated in Figure 4). Data were obtained on a standard phantom (Gammex Sono410 SCG, schema illustrated Figure 2b) with a cyst and a "grappe" and the orthogonal $(x, z)$ (a) and $(y, z)$ (b) planes are shown. The planes are taken in the center of the transducer's array, at $x = 0$ mm or $y = 0$ mm, respectively. 9 PW were used for the acquisition illustrated in the first row, while 9 DW were used for the acquisition shown in the lower row.

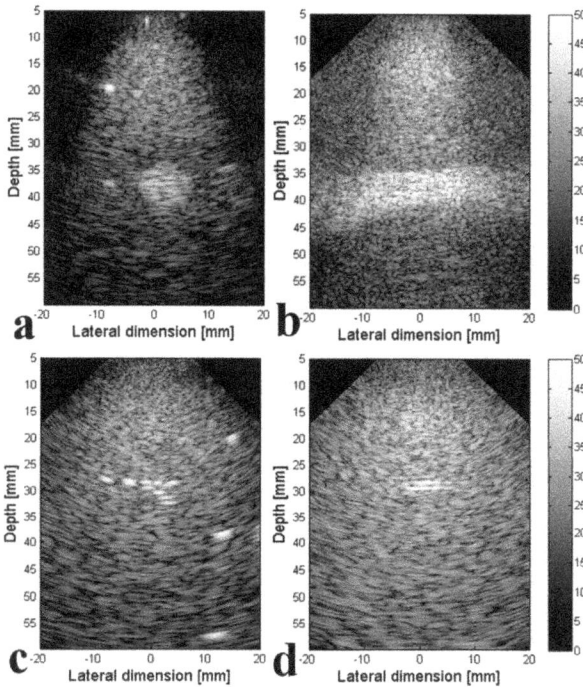

**Figure 4.** Experimental data obtained for the Gammex phantom with 1024 active elements in $(x, z)$ (**a,c**) and $(y, z)$ (**b,d**) planes on a cyst (**a,b**) by using 9 PW and "grappe" wires (**c,d**) by using 9 DW. A dynamic of 50 dB has been used in all images.

Different configurations of the US matrix array, by using 256, 512 or 1024 elements, were also investigated using the 4-D US platfom described above. Figure 5 illustrates an example of volumetric acquisition on a standard phantom with several wires positioned at different depths. Nine DW were transmitted in each of the $x$ and $y$ dirrections for this example. The $(x, z)$ and $(y, z)$ planes are shown here, the planes being taken at the center of the active array. As we expected, after visual inspection, the experimental results obtained here clearly indicate the full active array of the US probe as the best configuration. At a rapid inspection, the $(x, z)$ plane indicates similar results, but at a closer view, some differences are identified. The wire positioned in the near field (depth = 10 mm) is very accurate with this configuration, while the configurations with 512 active elements still illustrate good accuracy. The 256-elements configuration (the last column) presents the worst case, as expected, taking into account the low number of active elements. The same observations are done if we inspect the wire situated in the far-field (depth = 50 mm), where the best accuracy is visible in the full active array configuration.

At the same time, the $(y, z)$ plane illustrates the degradation of the image quality with lower numbers of active elements. The step in the $y$ direction is modified according to the configuration used, this leading to apparition of important artefacts for some cases, in the near and far-field. Not all the lengths of the wires are visible in this plane, due to small misalignment between the US probe and the wires.

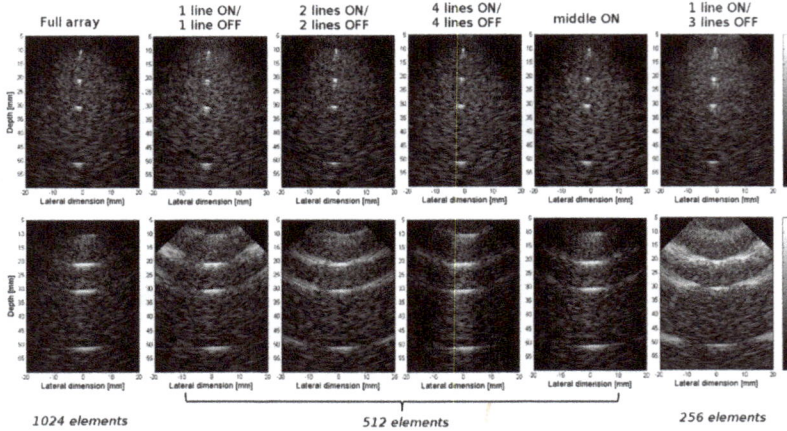

**Figure 5.** Experimental data acquired with 9 DW on a standard phantom. The $(x, z)$ plane (first row) and the $(y, z)$ plane (second row) are shown in the center of the US device. Different configurations of the array were used: full array, 1 line ON-1 line OFF, 2 lines ON-2 lines OFF, 4 lines ON-4 lines OFF, middle ON and 1 line ON-3 lines OFF. A dynamic of 50 dB has been used in all the images.

Further quantitative results of FWHM are measured and displayed in Table 4, for the $(x, z)$ plane, in axial and lateral direction. These results confirm the previous observations, indicating the degree of degradation of the image quality when downsampled configurations are used.

**Table 4.** FWHM calculated in the $(x, z)$ plane over three scatters located at depths 10, 20, 30 mm and the CR and CNR evaluated on the $(y, z)$ central slices of the acquired volumes.

| Resolution (mm) | | Full Array | 1 Line ON/1 Line OFF | 2 Lines ON/ 2 Lines OFF | 4 Lines ON/ 4 Lines OFF | Middle ON | 1 Line ON/ 3 Lines OFF |
|---|---|---|---|---|---|---|---|
| | axial 10 mm | 1.8 | 2.5 | 2.4 | 2.8 | 2.9 | 4 |
| | axial 20 mm | 1.7 | 2.5 | 3.1 | 2.2 | 2.3 | 3.3 |
| | axial 30 mm | 1.9 | 2.5 | 2 | 1.7 | 2.5 | 2.5 |
| | **Average** | **1.8** | **2.5** | **2.5** | **2.2** | **2.5** | **3.3** |
| $(x, z)$ plane | lateral 10 mm | 0.9 | 1.6 | 1.7 | 0.9 | 1 | 2.3 |
| | lateral 20 mm | 3 | 2.9 | 2.8 | 1.9 | 2.9 | 3.1 |
| | lateral 30 mm | 2 | 2.9 | 3 | 3.1 | 3 | 3.4 |
| | **Average** | **2** | **2.4** | **2.5** | **2** | **2.3** | **2.9** |
| **CR (dB)** | | Full Array | 1 line ON/1 line OFF | 2 lines ON/2 lines OFF | 4 lines ON/4 lines OFF | Middle ON | 1 line ON/3 lines OFF |
| $(y, z)$ plane | | 7.71 | 2.84 | 2.30 | 4.20 | 3.50 | 2.02 |
| **CNR (dB)** | | Full array | 1 line ON/1 line OFF | 2 lines ON/2 lines OFF | 4 lines ON/4 lines OFF | Middle ON | 1 line ON/3 lines OFF |
| $(y, z)$ plane | | 2.99 | −5.36 | −6.54 | −2.61 | −4.25 | −7.31 |

Another example of the 9 DW acquisition with the same configurations of the US transducer is shown in Figure 6. A "grappe" wire was imaged in this standard phantom. The same general observations can be done as in the previous example. Additionally, in the $(x, z)$ plane, the separation of 2 different wires positioned very close to each-other (at depth = 28 mm) is very hard to identify, except for the case of the full active array (first column). Moreover, the contour of the wires positioned at the border of the image, not directly in front of the US probe, are very hard to identify for configurations with a lower number of elements. Important secondary lobes and image degradation is visible in the $(y, z)$ planes illustrated here. The two parallel lines are clearly distinguished when using 1024 elements, but for 256 elements and even for some cases with 512 elements, it is hard to define them.

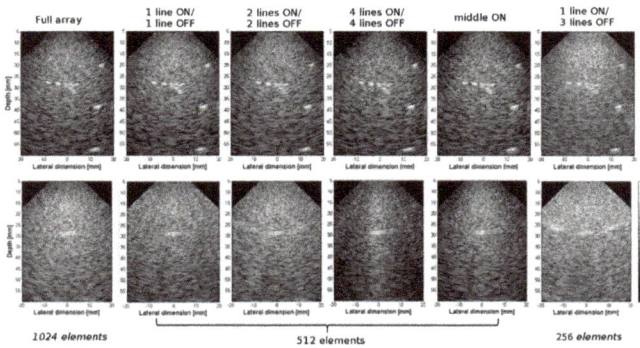

**Figure 6.** Experimental data acquired with 9 DW on a "grape" wire phantom. Different configurations of the US array were used: full array, 1 line ON-1 line OFF, 2 lines ON-2 lines OFF, 4 lines ON-4 lines OFF, middle ON and 1 line ON-3 lines OFF. The shown planes $(x, z)$ (first row) and $(y, z)$ (second row) are taken in the center of the US device. A dynamic of 50 dB has been used in all the images.

An example of 9 PW acquisition on a cyst phantom and wires placed at different depths is illustrated in Figure 7. In the $(x, z)$ plane, in the near field, we can identify the presented scatters (at depth = 7 mm and 19 mm) for all the cases, but in the far-field, the contour of the cyst is affected by lower number of elements compared with the full-matrix array. The anechoic cyst at 40 mm depth is clearly detectable with all the configurations tested, in both $(x, z)$ and $(y, z)$ planes but its borders are less visible in $(y, z)$ planes (second row) with downsampled configurations, the degradation of the obtained images being clearly visible for cases with 512 or 256 elements. The measured values for CR and CNR obtained for each of these acquisitions in $(x, z)$ central planes are quite close for all the configurations due to the presence of the same number of active elements in the $x$ direction, while the values obtained in the $(y, z)$ plane are displayed in Table 4, the last 2 rows. Both measurements indicate the best results for the full active configuration array, followed by the configuration 4 lines ON/4 lines OFF. The worst case when considering only the configurations with 512 active elements is represented by 2 lines ON/2 lines OFF, both for CR (a value of 2.3 dB versus 4.2 dB for 4 lines ON/4 lines OFF) and CNR (a value of $-6.54$ dB versus 2.61 dB for 4 lines ON/4 lines OFF). As expected, the worst results are obtained when using only 256 elements, due to the limited number of elements and important pitch between the active lines.

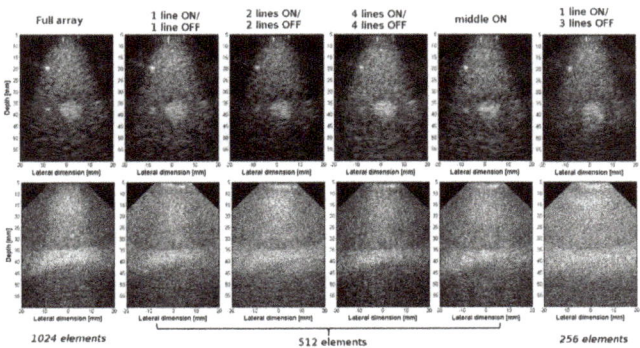

**Figure 7.** Experimental data acquired with 9 PW on a standard cyst phantom. Different configurations of the US array were used: full array, 1 line ON-1 line OFF, 2 lines ON-2 lines OFF, 4 lines ON-4 lines OFF, middle ON and 1 line ON-3 lines OFF. The shown planes, $(x, z)$ (first row) and $(y, z)$ (second row) are taken in the center of the US device. A dynamic of 50 dB has been used in all the images.

*3.3. Fast Volumetric Imaging for Doppler Application*

The feasibility of the 4-D platform implementation is shown here on a blood vessel mimicking phantom (described in Section 2.4) for 3-D Doppler applications. The results indicate the capability to extract the Power Doppler information when fast volumetric imaging is performed. Two different cases are presented in Figure 8 the US probe parallel (first row) and tilted by an angle of 12° from the vessel phantom (second row). One PW was transmitted for each acquisition. In B-mode, the walls of the phantom are easily detected in all the planes without any distortion. The frame-rate available for volumetric acquisitions is sufficient to detect the blood mimicking fluid for power Doppler evaluation. Some artifacts are visible in the presented cases, on the distant wall of the imaged phantom.

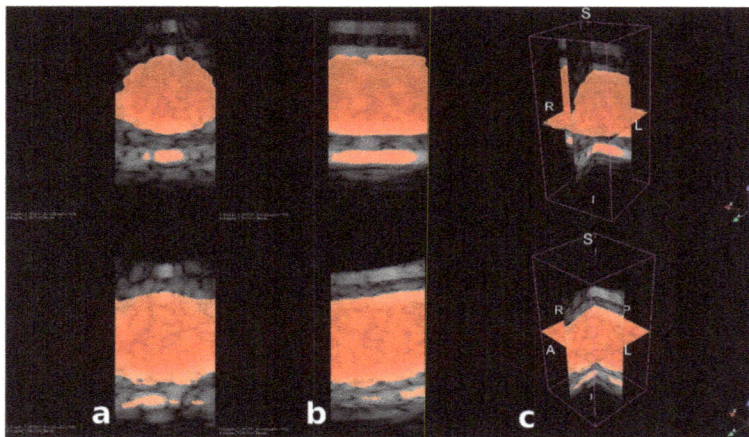

**Figure 8.** Examples of Power Doppler illustrations of two US acquisitions on a blood vessel mimicking phantom. The US probe is parallel ($\theta = 0$) to the phantom-tube for the first example (first row) while the probe was tilted at an angle of 12° from the phantom-tube. Axial (**a**), lateral (**b**) and 3-D view rendering (**c**) is illustrated here. A dynamic of 50 dB has been used for the B-mode grayscale image.

## 4. Discussion

This study illustrates the development and feasibility of a real-time 4-D US imaging platform in full array mode by synchronizing emission and reception of up to 1024 elements independently. The initial experiments conducted in this study demonstrate the flexibility of the modular US platform, having the possibility to reconfigure it according to the number of available Vantage systems (between 1 and 4). This allows us to use full array, dense/halves/quarters of the array or to downsample (regular or sparse) the full array, as needed. Different imaging strategies were investigated using this US platform, including plane, diverging or focused waves and also high frame-rate imaging techniques. Moreover, a Doppler detection strategy was implemented as a first application by using volumetric fast imaging.

Preliminary numerical simulations indicated the image quality for some cases with full and with regular downsampled array when using the $32 \times 32$ US matrix. This allowed us to understand how the image quality will change for different configurations and to determine the best physical addressing of the US probe final design. The experimental data obtained when using the same configurations of the transducer showed the feasibility of real-time 4D US acquisitions in simultaneous mode in different systems.

As expected, the best configuration in terms of image quality was obtained experimentally when the full matrix array was active while using 4 Vantage systems. The worst configuration included only 256 elements, with only one US system was required. At the same time, this is the only configuration allowing processing of the data and displaying the image in real time, this issue being very important for positioning and guidance. Future work will be focused on developing strategies for increasing the image quality when using one single system, in order to be able to provide images of clinical use, especially for moving organs. The implementation in synchronous mode will follow an increase in the number of active elements, but the data must be reprocessed offline. Knowing that 3-D volume images are acquired, the data offer great utility for offline analyses even after the examination is done, reducing repeat examination and making it more cost-effective for the patient.

The integration of the four 256-channels systems into a single 1024-channels system is theoretically possible, but requires further technical developments of the constructor. Such an evolution could facilitate the transition to clinical applications, opening the possibility to collect and process the data acquired by all 1024 channels in near real-time, improving the image quality.

Overall, when inspecting the $(x, z)$ plane positioned in the center of the transducer array, no major differences were detected due to the same number of active elements (32) in the $x$ direction with all the tested configurations. When using a larger pitch (in $y$ direction) for a reduced number of elements, more important secondary lobes are visible in the $(y, z)$ plane and therefore, we have lower image contrast. Depending on the pitch value, the secondary lobes appear more or less visible, illustrating the degradation of the images. The same result is obtained when using PW and DW.

Among the compared configurations with 512 elements, in the near-field, no visual difference is observed when using the same number of elements, while in the far-field and on the lateral parts of the images, the contours are seriously affected when using only the central part of the US probe (see Figures 5 and 7). In addition, when evaluating the FWHM, the configuration with 2 lines ON/2 lines OFF and 1 line ON/1 line OFF seems to be the closer to the 1024 elements, but the measured values for all these configurations remain quite close.

Moreover, a lower transmit energy compromises the image quality by diminishing the CR and CNR, from 7.71 dB (CR) and 2.99 dB (CNR), when using full active array, to 2.02 dB (CR) and −7.31 dB (CNR), respectively, for the configuration with 256 active elements. The results obtained experimentally are in good agreement with preliminary numerical simulations, and the image degradation is explained by the different values of the pitch in the $y$ direction.

By using the presented 4-D US platform, the power Doppler technique was exploited to detect moving particles. Here, only the full configuration array was used to prove the feasibility of the method, the experiments with lower number of elements being beyond the scope of this article. This method is particularly useful when examining superficial structures. The capability of obtaining entire volumes at a high frame-rate (used PRF acquisitions on blood vessel mimicking phantom was 2000 Hz) opens a wide field of vascular and cardiac applications. The feasibility of the ultrafast imaging is very important in the context of cardiac imaging. Here, the difficulty is the heart beating. Therefore, the high frame-rate allows performing the acquisition in less time in order to not be disturbed by heart movement [23].

The combination of hardware and software-based technologies provides full flexibility of the developed platform, its reconfgurability being one of its major advantages. This makes it suitable for different US applications requiring 1024 channels or less in regular or optimized sparse arrays. As the US scanners are completely open systems, our platform can be programmed to support different strategies which typically cannot be implemented in conventional data flow architectures. Different methods can be investigated: novel transmission and/or reception methods and sequences, apodization windowing [35], non-conventional beamforming techniques, excitation amplitude and aperture control, etc., but also custom-data processing. Such a platform will indeed be crucial to objectivize the real adjunct and efficiency of advanced reconstruction strategy, aiming at providing similar image quality as from highly undersampled acquisition since they can be compared against ground truth available for a full configuration. Indeed, over-performing methods are often published without a true reference implementation of the acquisition, and data are generally virtually undersampled, lacking realism.

*Appl. Sci.* **2018**, *8*, 200

The connectors of the US probe were chosen in order to allow adapting the impedance while for all 4 systems we have the possibility to have the HIFU (High Intensity Focused Ultrasound) option. Different studies have already illustrated the possibility to use HIFU for pre-clinical and clinical cardiac applications such as: treatment of arrhythmias [36,37], atrial septostomy [38] or atrio-ventricular nodal-ablation in dogs [39]. Therefore, the developed 4-D US platform can be considered as dual-mode, imaging and therapy. This aspect can be exploited, even using low intensity therapies (the US probe presented in this manuscript is not developed for HIFU mode), for theragnostic applications in the heart, for example. Here, the major interest is represented by the use of the same system for imaging and therapy.

One of the technical limitations of this platform when using the described US probe involves the physical addressing of the elements. When using 512 active elements in configurations "middle ON" and "4 lines OFF/4 lines ON", we still need 4 synchronized US systems even if not all the channels are used. This issue could be improved by using US probes with a different addressing connectivity.

The next technical step includes technical improvements for near real-time visualization of data received by several systems: high computational power and storage capability will be required. Therefore, the large acquired data flow will be rapidly transferred, stored and processed, the complementary information provided by the 4 systems being visualized immediately.

By using this powerful US platform, the advantages offered by a 2-D probe with a high number of channels can be easily investigated. Moreover, the flexibility of the system allows for implementing a large class of US applications, in therapy or monitoring. The technical feasibility of real-time 4-D low energy US for imaging and potentially for therapeutic purposes was confirmed with this 1024-channels high-density US system, offering full research access for developing advanced US strategies for different applications.

## 5. Conclusions

A new powerful high-channels US platform for implementing 4-D (real-time 3-D) advanced US strategies, offering full research access, was developed and presented here. The conducted experiments demonstrate its technical feasibility and flexibility, allowing us to use a full or downsampled array, as needed, illustrating the degree of degradation of image quality for each combination. Different imaging strategies were investigated using this US platform, and a Doppler detection strategy was implemented as a first application by using volumetric fast imaging, opening a large field of cardiac and vascular applications, in therapy or monitoring.

**Supplementary Materials:** The following are available online at www.mdpi.com/2076-3417/8/2/200/s1. The following Supplementary Materials are available online: Animation of the 3-D rendering of Power Doppler obtained when the US probe is parallel to the vessel phantom, corresponding to the example illustrated in Figure 8, first row. Animation of the 3-D rendering of Power Doppler obtained when the US probe is tilted from the vessel phantom corresponding to the example illustrated in Figure 8, second row.

**Acknowledgments:** This work was completed as part of the framework of the "Programme Avenir Lyon Saint-Etienne" of the Université de Lyon (ANR-11-IDEX-0007), within the Program "Investissements d'Avenir" operated by the French National Research Agency and co-founded by the People Programme (Marie Curie Actions) of the European Union's Seventh Framework Programme (FP7/2007-2013) under REA grant agreement n° PCOFUND-GA-2013-609102, through the PRESTIGE programme coordinated by Campus France. It was also performed within the framework of the Labex PRIMES (ANR-10-LABX-0063) of Université de Lyon, within the program "Investissements d'Avenir" (ANR-11-IDEX-0007) operated by the French Nation-al Research Agency (ANR). 2 Verasonics systems were cofounded by the FEDER program, Saint-Etienne Metropole (SME) and Conseil General de la Loire (CG42) within the framework of the SonoCardioProtection Project leaded by Pr Pierre Croisille. The authors would also like to thank LabTAU for their contribution to the development of the 32 × 32 probe prototype compatible with driving 1 to 4 systems as well as for the provision of the probe and two Vantage 256 systems.

**Author Contributions:** L.P., F.V. W.A.N. designed and carried out the simulations and experimental studies; L.P. and F.V. analyzed the data; R.S. and A.B. provided technical support; J.Y.C., H.L. W.A.N. and M.V. provided critical feedback for the different steps of this project ; L.P. wrote the original draft. All co-authors contributed with reading and improving the final manuscript.

**Conflicts of Interest:** The authors declare no conflict of interest.

## References

1. Tortoli, P.; Bassi, L.; Boni, E.; Dallai, A.; Guidi, F.; Ricci, S. ULA-OP: An advanced open platform for ultrasound research. *IEEE Trans. Ultrason. Ferroelectr. Freq. Control* **2009**, *56*, 2207–2216. [CrossRef]
2. Asef, A.A.; Maia, J.M.; Costa, E.T. A flexible multichannel FPGA and PC-Based ultrasound system for medical imaging research: Initial phantom experiments. *Res. Biomed. Eng.* **2015**, *31*, 277–281.
3. Brunke, S.S.; Insana, M.F.; Dahl, J.J.; Hansen, C.; Ashfaq, M.; Ermert, H. An ultrasound research interface for a clinical system. *IEEE Trans. Ultrason. Ferroelectr. Freq. Control* **2007**, *54*, 198–210. [CrossRef]
4. Roux, E.; Varray, F.; Petrusca, L.; Cachard, C.; Tortoli, P.; Liebgott, H. 3D diverging waves with 2D sparse arrays: A feasibility study. In Proceedings of the IEEE International Ultrasonics Symposium, Washington, DC, USA, 6–9 September 2017.
5. Roux, E.; Ramalli, A.; Tortoli, P.; Cachard, C.; Robini, M.; Liebgott, H. 2D ultrasound sparse arrays multi-depth radiation optimization using simulated annealing and spiral-array inspired energy functions. *IEEE Trans. Ultrason. Ferroelectr. Freq. Control* **2016**, *63*, 2138–2149.
6. Badescu, E.; Bujoreanu, D.; Petrusca, L.; Friboulet, D.; Liebgott, H. Multi-line transmission for 3D ultrasound imaging: An experimental study. In Proceedings of the IEEE International Ultrasonics Symposium, Washington, DC, USA, 6–9 September 2017.
7. Zhang, M.; Varray, F.; Besson, A.; Carrillo, R.E.; Viallon, M.; Garcia, D.; Thiran, J.P.; Friboulet, D.; Liebgott, H.; Bernard, O. Extension of fourier-based techniques for ultrafast imaging in ultrasound with diverging waves. *IEEE Trans. Ultrason. Ferroelectr. Freq. Control* **2016**, *63*, 2125–2137. [CrossRef]
8. Salles, S.; Liebgott, H.; Garcia, D.; Vray, D. Full 3D transverse oscillations: A method for tissue motion estimation. *IEEE Trans. Ultrason. Ferroelectr. Freq. Control* **2015**, *62*, 1473–1485. [CrossRef]
9. Tanter, M.; Fink, M. Ultrafast imaging in biomedical ultrasound. *IEEE Trans. Ultrason. Ferroelectr. Freq. Control* **2014**, *61*, 102–119. [CrossRef]
10. Synnevag, J.-F.; Austeng, A.; Holm, G. Adaptive beamforming applied to medical ultrasound imaging. *IEEE Trans. Ultrason. Ferroelectr. Freq. Control* **2007**, *54*, 160–1613. [CrossRef]
11. Provost, J.; Papadacci, C.; Arango, J.E.; Imbault, M.; Fink, M.; Gennisson, J.L.; Tanter, M.; Pernot, M. 3D ultrafast ultrasound imaging in vivo. *Phys. Med. Biol.* **2014**, *59*, L1–L13. [CrossRef]
12. Jensen, J.A.; Holten-Lund, H.; Nilsson, R.T.; Hansen, M.; Larsen, U.D.; Domsten, R.P.; Tomov, B.G.; Stuart, M.B.; Nikolov, S.I.; Pihl, M.J.; et al. SARUS: A synthetic aperture real-time ultrasound system. *IEEE Trans. Ultrason. Ferroelectr. Freq. Control* **2013**, *60*, 1838–1852. [CrossRef]
13. Provost, J.; Papadacci, C.; Demene, C.; Gennisson, J.L.; Tanter, M.; Pernot, M. 3-D ultrafast doppler imaging applied to the noninvasive mapping of blood vessels in vivo. *IEEE Trans. Ultrason. Ferroelectr. Freq. Control* **2015**, *62*, 1467–1472. [CrossRef] [PubMed]
14. Papadacci, C.; Finel, V.; Provost, J.; Villemain, O.; Bruneval, P.; Gennisson, J.L.; Tanter, M.; Fink, M.; Pernot, M. Imaging the dynamics of cardiac fiber orientation in vivo using 3D Ultrasound Backscatter Tensor Imaging. *Sci. Rep.* **2017**, *7*, 830. [CrossRef] [PubMed]
15. Savord, B.; Solomon, R. Fully sampled matrix transducer for real time 3D ultrasonic imaging. In Proceedings of the IEEE Symposium on Ultrasonics, Honolulu, HI, USA, 5–8 October 2003; Volume 1, pp. 945–953.
16. Bhuyan, A.; Choe, J.W.; Lee, B.C.; Wygant, I.O.; Nikoozadeh, A.; Oralkan, Ö.; Khuri-Yakub, B.T. Integrated circuits for volumetric ultrasound imaging with 2-D CMUT arrays. *IEEE Trans. Biomed. Circuits Syst.* **2013**, *7*, 796–804. [CrossRef] [PubMed]
17. Kortbek, J.; Jensen, J.A.; Gammelmark, K.L. Sequential beamforming for synthetic aperture imaging. *Ultrasonics* **2013**, *53*, 1–16. [CrossRef] [PubMed]
18. Christiansen, T.L.; Rasmussen, M.F.; Bagge, J.P.; Nordahl Moesner, F.; Jensen, J.A.; Thomsen, E.V. 3-D imaging using row-column-addressed arrays with integrated apodization—Part II: Transducer fabrication and experimental results. *IEEE Trans. Ultrason. Ferroelectr. Freq. Control* **2015**, *62*, 959–971. [CrossRef] [PubMed]
19. Didier, D.; Brusseau, E.; Detti, V.; Varray, F.; Basarab, A. Ultrasound medical imaging. In *Medical Imaging Based on Magnetic Fields and Ultrasounds*; John Wiley & Sons: Hoboken, NJ, USA, 2014; pp. 1–72.
20. Papadacci, C.; Pernot, M.; Couade, M.; Fink, M.; Tanter, M. High-contrast ultrafast imaging of the heart. *IEEE Trans. Ultrason. Ferroelectr. Freq. Control* **2014**, *61*, 288–301. [CrossRef] [PubMed]
21. Hasegawa, H.; Kanai, H. High-frame-rate echocardiography using diverging transmit beams and parallel receive beamforming. *J. Med. Ultrason.* **2011**, *38*, 129–140. [CrossRef]

22. Tong, L.; Ramalli, A.; Tortoli, P.; Fradella, G.; Caciolli, S.; Luo, J.; D'hooge, J. Wide-angle tissue doppler imaging at high frame rate using multi-line transmit beamforming: An experimental validation in vivo. *IEEE Trans. Med. Imaging* **2016**, *35*, 521–528. [CrossRef]

23. Poree, J.; Posada, D.; Hodzic, A.; Tournoux, F.; Cloutier, G.; Garcia, D. High-frame-rate echocardiography using coherent compounding with doppler-based motion-compensation. *IEEE Trans. Med. Imaging* **2016**, *35*, 1647–1657. [CrossRef]

24. Joos, P.; Liebgott, H.; Varray, F.; Petrusca, L.; Garcia, D.; Vray, D.; Nicolas, B. High-frame-rate 3-D echocardiography based on motion compensation: An in vitro evaluation. In Proceedings of the IEEE International Ultrasonics Symposium, Washington, DC, USA, 6–9 September 2017.

25. Cikes, M.; Tong, L.; Sutherland, G.R.; D'hooge, J. Ultrafast cardiac ultrasound imaging. *JACC Cardiovasc. Imaging* **2014**, *7*, 812–823. [CrossRef]

26. Vappou, J.; Luo, J.; Konofagou, E.E. Pulse wave imaging for noninvasive and quantitative measurement of arterial stiffness in vivo. *Am. J. Hypertens.* **2010**, *23*, 393–398. [CrossRef] [PubMed]

27. Bercoff, J.; Montaldo, G.; Loupas, T.; Savery, D.; Mézière, F.; Fink, M.; Tanter, M. Ultrafast compound Doppler imaging: Providing full blood flow characterization. *IEEE Trans. Ultrason. Ferroelectr. Freq. Control* **2011**, *58*, 134–147. [CrossRef] [PubMed]

28. Jensen, J.A. Field: A program for simulating ultrasound systems. In Proceedings of the 10th Nordic-Baltic Conference on Biomedical Imaging Published in Medical & Biological Engineering & Computing, Tampere, Finland, 9–13 June 1996; Volume 34, pp. 351–353.

29. Jensen, J.A.; Svendsen, N.B. Calculation of pressure fields from arbitrarily shaped, apodized, and excited ultrasound transducers. *IEEE Trans. Ultrason. Ferroelectr. Freq. Control* **1992**, *39*, 262–267. [CrossRef] [PubMed]

30. Perrot, V.; Petrusca, L.; Bernard, A.; Vray, D.; Liebgott, H. Simultaneous pulse wave and flow estimation at high-framerate using plane wave and transverse oscillation on carotid phantom. In Proceedings of the IEEE International Ultrasonics Symposium, Washington, DC, USA, 6–9 September 2017.

31. Fromageau, J.; Gennisson, J.L.; Schmitt, C.; Maurice, R.; Mongrain, R.; Cloutier, G. Estimation of polyvinyl alcohol cryogel mechanical properties with four ultrasound elastography methods and comparison with gold standard testings. *IEEE Trans. Ultrason. Ferroelectr. Freq. Control* **2007**, *54*, 498–509. [CrossRef] [PubMed]

32. Ramnarine, K.V.; Nassiri, D.K.; Hoskins, P.R.; Lubbers, J. Validation of a new blood-mimicking fluid for use in doppler flow test objects. *Ultrasound Med. Biol.* **1998**, *24*, 451–459. [CrossRef]

33. Montaldo, G.; Tanter, M.; Bercoff, J.; Benech, N.; Fink, M. Coherent plane-wave compounding for very high frame rate ultrasonography and transient elastography. *IEEE Trans. Ultrason. Ferroelectr. Freq. Control* **2009**, *56*, 489–506. [CrossRef] [PubMed]

34. Fedorov, A.; Beichel, A.; Kalpathy-Cramer, J.; Finet, J.; Fillion-Robin, J.C.; Pujol, S.; Bauer, C.; Jennings, D.; Fennessy, F.; Sonka, M.; et al. 3D slicer as an image computing platform for the Quantitative Imaging Network. *Magn. Reson. Imaging* **2012**, *30*, 1323–1341. [CrossRef] [PubMed]

35. Tomov, B.G.; Jensen, J.A. Compact implementation of dynamic receive apodization in ultrasound scanners. In *Medical Imaging 2004: Ultrasonic Imaging and Signal Processing*; SPIE: San Diego, CA, USA, 2004.

36. Ninet, J.; Roques, X.; Seitelberger, R.; Deville, C.; Pomar, J.L.; Robin, J.; Jegaden, O.; Wellens, F.; Wolner, E.; Vedrinne, C.; et al. Surgical ablation of atrial fibrillation with off-pump, epicardial, high-intensity focused ultrasound: Results of a multicenter trial. *J. Thorac. Cardiovasc. Surg.* **2005**, *130*, 803–809. [CrossRef] [PubMed]

37. Bessiere, F.; N'djin, W.A.; Colas, E.C.; Chavrier, F.; Greillier, P.; Chapelon, J.Y.; Chevalier, P.; Lafon, C. Ultrasound-guided transesophageal high-intensity focused ultrasound cardiac ablation in a beating heart: A pilot feasibility study in pigs. *Ultrasound Med. Biol.* **2016**, *42*, 1848–1861. [CrossRef] [PubMed]

38. Takei, Y.; Muratore, R.; Kalisz, A.; Okajima, K.; Fujimoto, K.; Hasegawa, T.; Arai, K.; Rekhtman, Y.; Berry, G.; Homma, S.; et al. In vitro atrial septal ablation using high-intensity focused ultrasound. *J. Am. Soc. Echocardiogr.* **2012**, *25*, 467–472. [CrossRef] [PubMed]

39. Strickberger, S.A.; Tokano, T.; Kluiwstra, J.U.; Morady, F.; Cain, C. Extracardiac ablation of the canine atrioventricular junction by use of high-intensity focused ultrasound. *Circulation* **1999**, *100*, 203–208. [CrossRef] [PubMed]

![applied sciences logo] *applied sciences*

MDPI

*Article*

# Contrast-Enhanced Ultrasound Imaging Based on Bubble Region Detection

Yurong Huang [1,†], Jinhua Yu [1,2,3,*], Yusheng Tong [3,4,†], Shuying Li [1], Liang Chen [3,4,*], Yuanyuan Wang [1,2] and Qi Zhang [5]

[1] Department of Electronic Engineering, Fudan University, Shanghai 200433, China; 15210720136@fudan.edu.cn (Y.H.); 14210720154@fudan.edu.cn (S.L.); yywang@fudan.edu.cn (Y.W.)
[2] Key Laboratory of Medical Imaging Computing and Computer Assisted Intervention of Shanghai, Shanghai 200433, China
[3] Institute of Functional and Molecular Medical Imaging, Fudan University, Shanghai 200030, China; ystong16@fudan.edu.cn
[4] Department of Neurosurgery, Huashan Hospital, Fudan University, Shanghai 200030, China
[5] School of Communication and Information Engineering, Shanghai University, Shanghai 200444, China; zhangq@shu.edu.cn
* Correspondence: jhyu@fudan.edu.cn (J.Y.); clclcl95@sina.com (L.C.); Tel.: +86-21-65643202 (J.Y.)
† Those authors contributed equally to this work.

Received: 4 September 2017; Accepted: 18 October 2017; Published: 24 October 2017

**Abstract:** The study of ultrasound contrast agent imaging (USCAI) based on plane waves has recently attracted increasing attention. A series of USCAI techniques have been developed to improve the imaging quality. Most of the existing methods enhance the contrast-to-tissue ratio (CTR) using the time-frequency spectrum differences between the tissue and ultrasound contrast agent (UCA) region. In this paper, a new USCAI method based on bubble region detection was proposed, in which the frequency difference as well as the dissimilarity of tissue and UCA in the spatial domain was taken into account. A bubble wavelet based on the Doinikov model was firstly constructed. Bubble wavelet transformation (BWT) was then applied to strengthen the UCA region and weaken the tissue region. The bubble region was thereafter detected by using the combination of eigenvalue and eigenspace-based coherence factor (ESBCF). The phantom and rabbit in vivo experiment results suggested that our method was capable of suppressing the background interference and strengthening the information of UCA. For the phantom experiment, the imaging CTR was improved by 10.1 dB compared with plane wave imaging based on delay-and-sum (DAS) and by 4.2 dB over imaging based on BWT on average. Furthermore, for the rabbit kidney experiment, the corresponding improvements were 18.0 dB and 3.4 dB, respectively.

**Keywords:** ultrasound contrast agent imaging; bubble wavelet transform; eigenspace; coherence factor

## 1. Introduction

Ultrasound contrast agents (UCAs) [1,2] are a type of diagnostic reagents that typically consist of gas-filled microbubbles with a diameter ranging from 1 to 10 µm. The microbubbles are filled with low solubility gas and are coated with a shell to prevent the microbubble from dissolving. UCAs are injected intravenously into the body and are considered safe for use in humans.

UCAs have been used clinically since the 1980s [1]. Ultrasound contrast agent imaging (USCAI) [3] has generated increased attention in recent years. As a result of the compressibility of microbubbles and the large acoustic impedance difference between them and the surrounding tissue, USCAI can greatly improve the contrast (CR) and contrast-to-tissue ratio (CTR) of a clinical ultrasound image.

*Appl. Sci.* **2017**, *7*, 1098; doi:10.3390/app7101098
147

The detection rate, sensitivity, and specificity with which small lesions can be detected can therefore be greatly increased [4,5].

Plane wave imaging (PWI) [6] has a great advantage over traditional B-mode imaging for USCAI. First, PWI can significantly reduce the destruction of microbubbles due to its low mechanical index. Second, PWI can track fast-moving microbubbles due to its high frame rate. However, poor imaging quality limits the clinical application of PWI.

A large proportion of novel USCAI technologies [7,8] appears to focus on the improvement of PWI imaging quality. Most of these existing methods enhance the image CTR by utilizing the abundant harmonic signals produced by UCAs. Generally speaking, the mainstream USCAI technologies can be classified into two categories: pulse coding [9] and bubble wavelet transformation (BWT) [10,11].

Pulse coding focuses on the transmitting end as it either changes the number of the transmit pulse, transmit phase, transmit frequency, or transmit amplitude. Pulse inversion (PI) [12], amplitude modulation (AM) [13], chirp encoded excitation [14], and Golay\encoded excitation [15] are typical representatives of such methods. Since the response of UCA is represented as harmonic signals while that of tissue represents predominantly fundamental signals, the tissue signal can be removed by PI and AM technology. Chirp-encoded excitation emits a long sequence with the frequency variation over time. Studies have shown that chirp excitation has the ability to strengthen the harmonic signals and lower the sub-harmonic generation threshold [16]. However, the matched filter is needed in the receiving end to decode the signal, which adds complexity.

Bubble wavelet transformation (BWT) is a novel type of USCAI technology proposed by Wan et al. [10,11]. BWT is utilized to analyze the correlation between the mother bubble wavelet and the received radio frequency (RF) signals. The constructed bubble wavelet obtained by simulating the microbubble model was highly correlated with the signals of microbubbles and has few similarities with the signals from surrounding tissues. The ability of BWT to enhance the imaging CTR has been validated by in vivo experiments. A mother wavelet was constructed by microbubble model simulation. The RF signals were then processed by the continuous wavelet transformation and a series of coefficient matrices was obtained. A better image quality and an enhancement of 6.0 dB in CTR can be obtained with BWT [11] while ensuring the high frame rate of PWI.

Both the pulse coding and BWT improved the CTR in the time-frequency domain. Under a low mechanical index, tissue produces a predominantly linear response and, thus, it is feasible to distinguish tissue and UCA according to their frequency components. The non-linear distortion of waveforms and the spectrum overlap between the tissue and UCA is an unfavorable factor.

In this paper, we take the tissue and UCA differences in both the frequency and spatial domains into consideration. PWI was used to improve the imaging frame rate and ensure the stability of UCA. A bubble wavelet based on the Doinikov model was firstly constructed. BWT was then applied to strengthen the UCA region and weaken the tissue region. Following this, the bubble region was subsequently detected by utilizing the combination of eigenvalues and eigenspace-based coherence factor (ESBCF). The residual tissue signal could be further suppressed, and a higher CTR was obtained.

## 2. Materials and Methods

### 2.1. Frequency Domain

#### 2.1.1. Bubble Wavelet Transformation (BWT)

With the rapid development of ultrasound molecular imaging, about a dozen of microbubble models [17] have been proposed. The dynamic behavior of a single microbubble under different ultrasound fields can be predicted with the help of these models. BWT applied the microbubble model to USCAI, where a novel mother wavelet named as a bubble wavelet was constructed based on the simulation results of the microbubble model.

The expression of BWT in the time domain can be described as:

$$bwt[x(t)] = \frac{1}{\sqrt{a}} \int_{-\infty}^{+\infty} x(t)\phi^*(\frac{t-b}{a})dt \tag{1}$$

where $x(t)$ is the signal to be processed; $\phi(t)$ is the bubble wavelet; $\phi(\frac{t-b}{a})$ is the function of the bubble wavelet after translation and scaling; $a$ is the scale factor; $b$ is the time-shifting factor; and superscript * denotes the conjugation operation.

In fact, BWT is the convolution operation of the bubble wavelet under different scale factors with the signal to be processed. The result of BWT is a series of wavelet coefficients. These coefficients, namely the function of the mother wavelet and the scale factor, illustrate the correlation between the bubble wavelet under a certain scale and the received signal. Due to the similarity of the frequency spectrum of the bubble wavelet and the UCA echoes, as well as the dissimilarity between this and the tissue echoes, the tissue signal can be weakened and the UCA signal can be strengthened after BWT.

In the application of BWT, the choice of the bubble wavelet and the scale factor are two key elements [18,19] that determine the image quality to a large extent. A more significant effect of CTR improvement can be achieved if the spectrum of the bubble wavelet is highly matched with that of UCA. The scale factor is another key issue. The scale resulting in the highest CTR is selected as the optimal factor.

### 2.1.2. Construction of Bubble Wavelet

Among the existing models, the Doinikov model [20] focuses on microbubbles with phospholipid shells. The Doinikov model has been proven to predict the 'compression-only' behavior [21] of Sonovue very well. It is commonly accepted that its spectrum is closer to that of the Sonovue microbubble and is the optimal choice for the bubble wavelet [17,22].

The Doinikov model can be described as:

$$\rho_l(RR'' + \tfrac{3}{2}R'^2) = (p_0 + \tfrac{2\sigma(R_0)}{R_0})(\tfrac{R_0}{R})^{3\gamma} - \tfrac{2\sigma(R_0)}{R} - 4\chi(\tfrac{1}{R_0} - \tfrac{1}{R}) - P_0 - P_{drive}(t) - 4\eta_l\tfrac{R'}{R} - 4(\tfrac{k_0}{1+\alpha|\frac{R'}{R}|} + \kappa_1\tfrac{R'}{R})\tfrac{R'}{R^2} \tag{2}$$

where $\rho_l = 1000$ kg/m³ denotes the density of the surrounding liquid; $P_0 = 101,000$ Pa as the atmospheric pressure; $\gamma = 1.07$ as the gas thermal insulation coefficient; $R_0 = 1.7$ μm as the initial radius of microbubble; $R$ is the instantaneous radius of microbubble; $R'$ is the first-order time derivative of $R$, with essentially $R' = dR/dt$ and $R'' = d^2R/dt^2$; $\sigma(R_0) = 0.072$ N/m as the initial surface tension; $\chi = 0.25$ N/m as the shell elasticity modulus; $\eta_l = 0.002$ PaS as the liquid viscosity coefficient; $k_0 = 4 \times 10^{-8}$ kg and $k_1 = 7 \times 10^{-15}$ kg/s as the shell viscosity components; $\alpha = 4$ μs as a characteristic time constant; and $P_{drive}(t)$ is the driving ultrasound.

The pressure scattered by the microbubble can be expressed as:

$$P(d) = \rho_L \frac{R}{d}(2R'^2 + RR'') \tag{3}$$

where $d$ denotes the distance from the center of the microbubble to the transducer.

Following this, the bubble wavelet can be obtained by solving Equations (2) and (3) with the initial condition of $R(t = 0) = R_0$, $R'(t = 0) = 0$.

### 2.2. Spatial Domain

Beamforming is an important part of the medical imaging system, and plays an important role in the imaging performance. Delay and sum (DAS) is the basic beamformer with a fixed weight. Its poor image quality promotes the development of adaptive beamformer. The minimum variance (MV) algorithm [23] is the originator of the adaptive beamformer. Eigenspace-based minimum variance (ESBMV) [24] was proposed since the improvement of MV in CR is not obvious.

Based on the observation of the differences between the UCA and the tissue region on the maximum eigenvalues and eigenspace-based coherence factor (ESBCF) index, we proposed a bubble region detection scheme based on the combination of eigenvalue and ESBCF index.

2.2.1. Eigenvalue

In the bubble region detection algorithm, the signal intensity is the most important characteristic. In the theory of ESBMV, the maximum eigenvalue belongs to the mainlobe signals [24]. The signal subspace is comprised of eigenvectors corresponding to the largest few eigenvalues, and the noise subspace is constructed by eigenvectors according to lower eigenvalues. The small eigenvalues tend to represent the noise field and the large eigenvalues represent the signal field [25]. Furthermore, the maximum eigenvalue gives an estimation of the signal power. Zeng et al. proposed a method to detect the signal based on the eigenvalues [26–28], which proves that it is practicable to distinguish the UCA and tissue region according to their eigenvalues from the side. Based on the observation of the experiment, we found that the amplitude of the maximum eigenvalue in the UCA region is obviously higher than that in the tissue region.

The eigenvalues are calculated as follows:

The RF signals were divided into several overlapping subarrays. The covariance matrix was computed after spatial smoothing and diagonal loading as follows:

$$R(k) = \frac{1}{M-L+1} \sum_{p=1}^{M-L+1} x_d{}^p(k) x_d{}^p(k)^H \qquad (4)$$

where $R$ is the covariance matrix; $k$ is the time index; $M$ is the transducer elements; $M-L+1$ denotes the overlapping subarray number; $L$ is the length of subarray; $p$ is the subarray number; $x_d{}^p(k)$ is the signal after delay; and $(\cdot)^H$ is the transposition.

The eigen-decomposition of the covariance matrix was required:

$$R = U\Lambda U^H \qquad (5)$$

where $\Lambda = diag(\lambda_1, \lambda_2, \ldots \lambda_L)$ is the square matrix of eigenvalues in the descending order and $U$ is the eigenvector corresponding to the eigenvalue.

By taking a phantom study as an example, the bubble detection ability of the eigenvalue is explained. The scheme of the phantom experiment is shown in Figure 1. The experiment system includes an ultrasonic research platform Verasonics Vantage 128 (Verasonics, Inc., Kirkland, WA, USA), a linear array transducer (L11-4v), a homemade gelatin phantom, a medical syringe, a tube, a beaker, a computer, Sonovue microbubble (Bracco Suisse SA, Switzerland), and fresh pork.

The phantom is 11 cm in length, 11 cm in width, 6 cm in height, and is made of gelatin. The process of making the phantom is as follows: (1) mix 50 g gelatin granules with 700 mL purified water; (2) heat the mixture for about 10 min until the gelatin particles are completely dissolved in water; (3) glue a plastic tube (whose diameter is 0.5 mm) to the sides of the plastic mold; (4) pour the mixture into the mold; (5) refrigerate the mixture in the freezer for about 3 h; (5) remove the mixture from the mold after the mixture is solidified; (6) draw the plastic pipe out of the phantom. A phantom with a wall-less tube is completed.

The fresh pork (the belly of a pig, 15 mm in thickness, 40 mm in length, and 25 mm in width) was bought from a local meat market on the day of the experiment. Verasonics was used to excite the ultrasound wave and record the RF data. Matlab was used to analyze the RF signal offline. The flowing Sonovue solution (diluted by 1000 times with 0.9% physiological saline) was injected into the pipe by a medical syringe, and the used solution flowed into the beaker. We placed a piece of fresh pork over the phantom. The ultrasonic coupling agent was applied between the pork and the phantom to ensure the signal transmission.

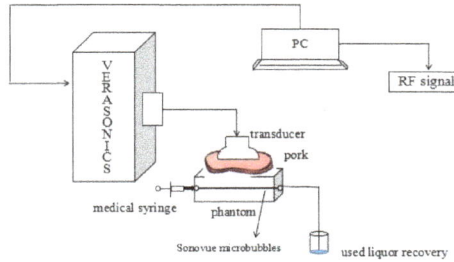

**Figure 1.** The phantom experiment setup (with pork).

The phantom image beamformed with DAS is shown in Figure 2, where the depth ranging from 0–15 mm is the pork area and that ranging from 30.5–35.5 mm is the bubble area. Taking the area at the width of 15 mm as an example, the maximum eigenvalue curve under different depths is shown in Figure 3. The area enclosed with the red rectangle represents the UCA region. Its maximum eigenvalue is quite large due to the existence of strong scattering signals produced by the UCA. The two purple ellipse areas are the interference region from tissue. For the artefact region, its eigenvalues are partially overlapped with the UCA due to the interference of the strong scattering signals from the UCA. On the other hand, in the pork part, its maximum eigenvalue is much smaller. Hence, we are able to eliminate the pork section preliminarily by setting an eigenvalue threshold. However, the disturbance of the artefact still exists. Thus, we further consider the application of ESBCF to remove the residual disturbance.

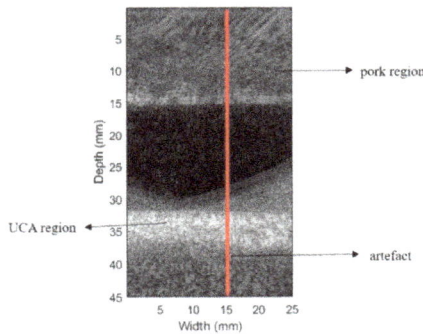

**Figure 2.** The image of the phantom (with pork) under delay-and-sum (DAS).

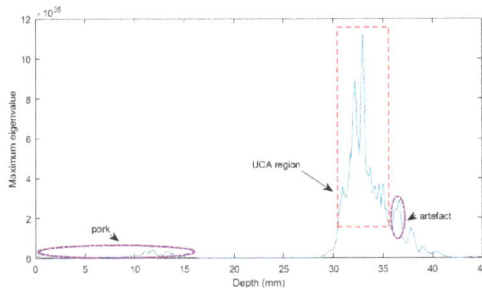

**Figure 3.** The maximum eigenvalue curve of different depths.

### 2.2.2. ESBCF

Coherence factor (CF) [29] is a type of adaptive weighting method based on the spatial spectrum of array data. In the last several years, a series of improved methods based on CF has been proposed [30]. ESBCF, proposed by Guo et al. [31], combined CF with a covariance matrix. The ability of ESBCF to detect point targets was verified by phantom and simulation experiments.

The description of ESBCF is as follows:

$$ESBCF_{vi}(\mathrm{k}) = \frac{(\frac{1}{M} \sum\limits_{n=1}^{M} v_{in}(k))^2}{\frac{1}{M} \sum\limits_{n=1}^{M} v_{in}(k)^2} = \frac{\sum (R_{vi})_{ele}}{M \cdot \sum (diag(R_{vi}))_{ele}} \tag{6}$$

where $R_{vi} = Vi^*Vi^H$ is the covariance matrix of eigenvector $Vi$; $i$ is the eigenvector index; and $n$ is the element index.

The maximum ESBCF value index curve of each imaging point is shown in Figure 4. For the UCA region, its maximum ESBCF value appears in the first few eigenvectors. In contrast, for the residual upper disturbing section, its maximum ESBCF value appears later. In this present study, the remaining upper interference can be removed by setting a maximum ESBCF index threshold.

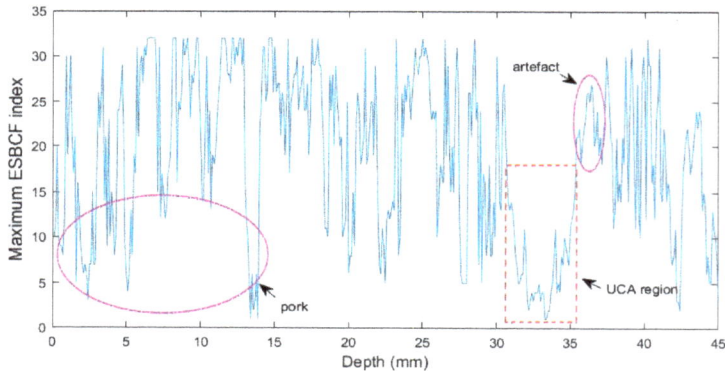

**Figure 4.** The maximum eigenspace-based coherence factor (ESBCF) value index curve of different depths.

### 2.2.3. Implementation of Bubble Region Detection

The calculation procedure of the bubble region detection method can be summarized as follows:

(1) BWT was first conducted to enhance the UCA signal, and the received RF signal was replaced by the wavelet coefficients under the optimal scale factor.
(2) The covariance matrix was computed according to Equation (4).
(3) The MV weight was obtained by minimizing the array noise output energy, which can be expressed as:

$$W_{MV} = \frac{R^{-1}d}{d^H R^{-1}d} \tag{7}$$

where $d$ is the direction vector; and $w$ is the weight vector.

(4) The eigen-decomposition of the covariance matrix was required:

$$R = U\Lambda U^H = U_S \Lambda_S U_S^H + U_P \Lambda_P U_P^H = R_S + R_P \tag{8}$$

where subscript $S$ denotes the signal and subscript $P$ is the noise subspace. The signal subspace is comprised of eigenvectors corresponding to the largest few eigenvalues ($\alpha$ times greater than $\lambda_1$ or $\beta$ times greater than $\lambda_L$).

(5) Based on the last step, the maximum eigenvalue and ESBCF for each imaging point was obtained. Following that, we found the maximum ESBCF index.

(6) We set the maximum eigenvalue and ESBCF index threshold to determine whether it is the UCA region or not.

(7) The ESBMV weight comes to:

$$W_{ESBMV} = U_{SU}U_{SU}^H W_{MV} \tag{9}$$

where $U_{SU}$ is the eigenvectors of the signal subspace of the detected UCA region.

(8) The region detection output is given by:

$$S(k) = \frac{1}{M-L+1} \sum_{p=1}^{M-L+1} W_{ESBMV}^H x_d^p(k) \tag{10}$$

(9) The final image can be formed after the signals are enveloped and the logarithmic transformation is applied.

### 2.2.4. Selection of Optimal Index

Scale Factor

By extending the time-domain expression of BWT to the frequency domain, Equation (1) changes into:

$$F\{cwt[x(t)]\} = F\{\frac{1}{\sqrt{a}} \int_{-\infty}^{+\infty} x(t)\phi^*(\frac{t-b}{a})dt\} = \frac{1}{\sqrt{a}}F[x(t)] \cdot F[\phi(\frac{t-b}{a})] = \sqrt{a}F(\omega)\psi(a\omega) \cdot e^{-j\omega b} \tag{11}$$

where $F$ denotes the Fourier transform.

In essence, BWT is a set of multi-scale filters that can control the passband by changing the scale factor. The correlation between the scale factor and the central frequency of the passband is as follows:

$$F(a) = \frac{F_C}{a} \tag{12}$$

where $F(a)$ is the central frequency corresponding to the scale factor $a$ and $Fc$ is the initial center frequency of the mother wavelet.

Theoretically, the optimal scale factor should be selected where the corresponding center frequency falls at the second harmonic of the UCA. We verified this hypothesis by traversing each scale factor. The optimal scale factor with different bubble wavelets under different transmit frequencies is shown in Table 1.

**Table 1.** The central frequency of the bubble wavelet with different transmit frequencies and their optimal scale factors.

| Transmit Frequency (MHz) | Central Frequency of Wavelet (*Fc*) | Optimal Scale Factor (*a*) | Central Frequency after Bubble Wavelet Transformation (BWT) (*F(a)*; in MHz) |
|---|---|---|---|
| 3 | 0.9394 | 0.16 | 5.87 |
| 4 | 0.6133 | 0.07 | 8.76 |
| 5 | 0.7869 | 0.08 | 9.83 |
| 6.25 | 0.8750 | 0.07 | 12.5 |

Eigenvalue

As is shown in Figure 3, the maximum eigenvalue curve of the UCA region is similar to a Gaussian function. The expression of the maximum eigenvalue curve can be represented as:

$$y = c * exp[-(\frac{x - \mu}{\sigma})^2] \tag{13}$$

where $x$ is the depth parameter; $\mu$ is the mean value; $\sigma$ is the variance; and $c$ is the amplitude. The eigenvalue of the UCA region can be set as $[\mu + e*\sigma, \mu - e*\sigma]$, where $e$ is an adjustable parameter.

Taking the area at the width of 10 mm as an example, the eigenvalue curve and the fitted curve is shown in Figure 5, where $c$ = 5.4e35, $\mu$ = 32.3, $\sigma$ = 4.2, and $e$ is chosen as 1.

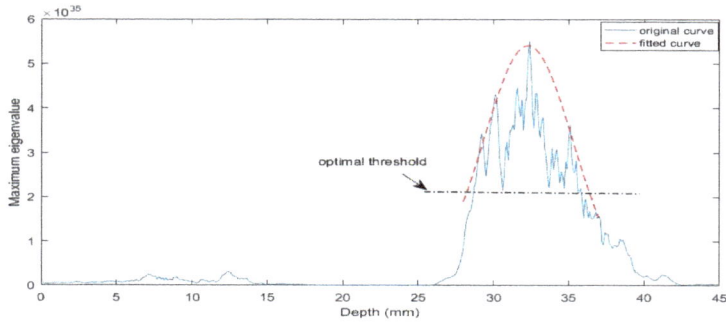

**Figure 5.** The fitted eigenvalue curve.

The ESBCF

Similarly, we fit the ESBCF curve by a Gaussian function. The ESBCF threshold can be set as $[\mu + e*\sigma, \mu - e*\sigma]$, where $e$ is an adjustable parameter. Taking the area at the width of 10 mm as an example, the eigenvalue curve and the fitted curve is shown in Figure 6.

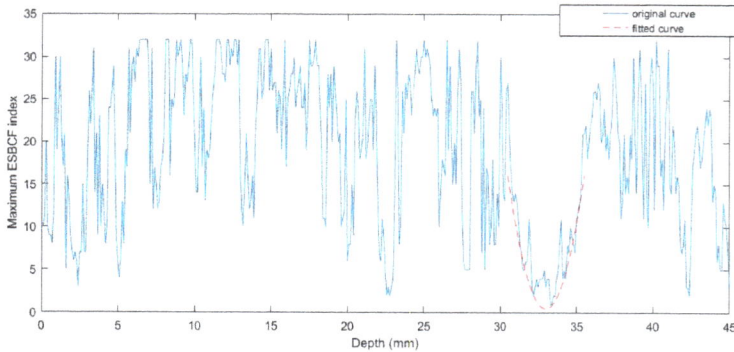

**Figure 6.** The fitted maximum ESBCF index curve.

## 3. Experiment and Results

### 3.1. Phantom Results

Two types of phantom experiments were designed. In the first, there were two wall-less pipes with different thicknesses in the phantom experiment (0.3 mm and 0.5 mm). In the other experiment, we placed a piece of fresh pork over the phantom, which only had one single pipe (about 0.5 mm). The detailed parameters for the phantom experiment are illustrated in Table 2.

**Table 2.** Parameters for the experiment.

| Experiment Parameters | Value |
|---|---|
| Transducer element number | 128 |
| Transducer element kerf | 0.05 mm |
| Transducer element width | 0.27 mm |
| Transducer element pitch | 0.3 mm |
| Transducer spacing between elements | 0. 295 mm |
| Transmit frequency | 3 MHz, 4 MHz, 5 MHz, 6.25 MHz |
| Transmit voltage | 1.6 V, 2.5 V, 5 V, 7.5 V, 10 V, 12.5 V, 15 V, 17.5 V, 20 V |
| Transmit pulse | Sine wave with two cycles |

Figure 7 provides the first phantom result (transmit frequency: 3 MHz, transmit voltage: 10 V). Figure 7a–c show the imaging using traditional DAS, MV, and ESBMV beamformer without BWT, respectively. Figure 7d is the image using BWT based on ESBMV beamformer and Figure 7e is the image with bubble region detection by setting the maximum eigenvalue and ESBCF index threshold based on Figure 7d. Some impurities in the phantom are removed and the two tubes filled with microbubbles are well-preserved using the proposed method.

Figure 8 provides the pork experiment results. As shown in Figure 8, DAS suffers from severe artefact interference, althouth the artefact interference was weakened with MV. The image CR can be further improved with ESBMV, although an obvious black region distortion appears inside the tube. The brightness of the UCA inside the tube became more uniform and UCA information loss could be solved with the help of BWT, although the interference of tissues still exists. In comparison, for bubble region detection, the region except for the microbubble was able to be fully eliminated.

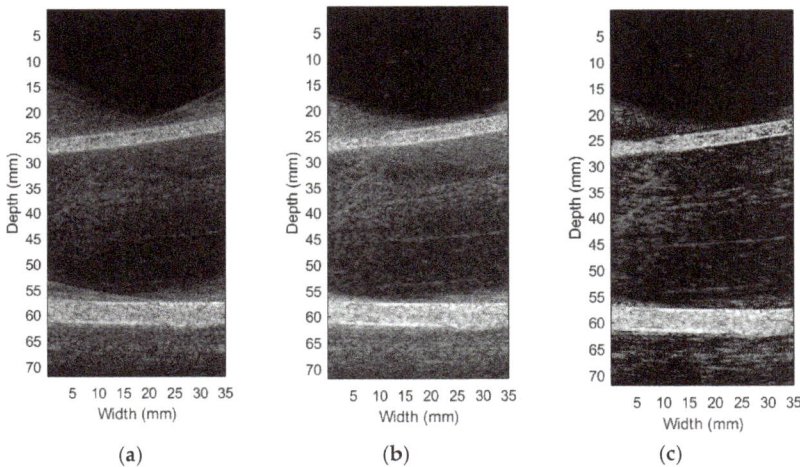

(a)   (b)   (c)

**Figure 7.** *Cont.*

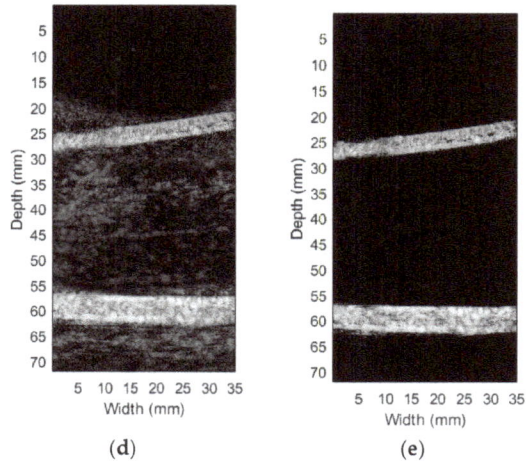

(d)  (e)

**Figure 7.** The image results of the phantom experiment (without pork): (**a**) DAS; (**b**) minimum variance (MV); (**c**) Eigenspace-based minimum variance (ESBMV); (**d**) BWT; (**e**) bubble region detection (dynamic range is 60 dB).

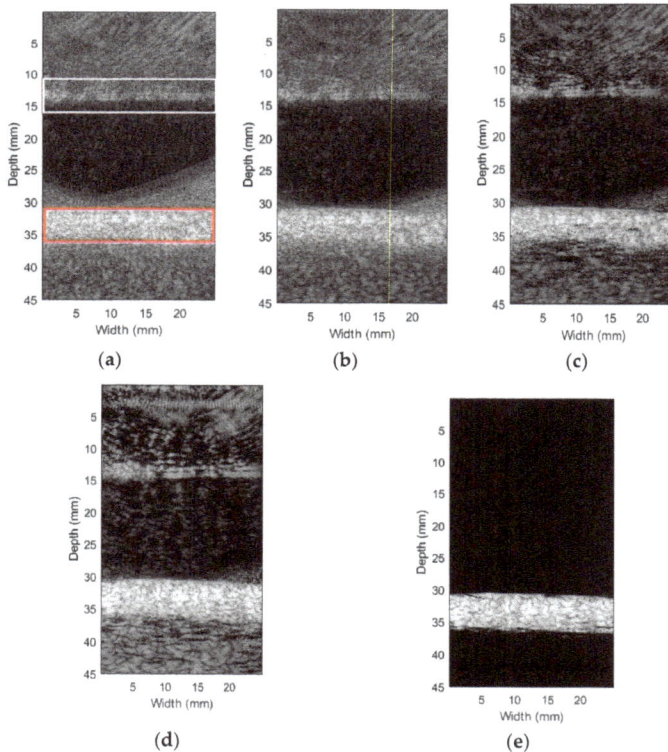

**Figure 8.** The image results of phantom experiment (with pork): (**a**) DAS; (**b**) MV; (**c**) ESBMV; (**d**) BWT; (**e**) bubble region detection.

CTR was calculated to describe the performance of different beamformers:

$$CTR = 20 \log \frac{I_{UCA}}{I_{tissue}} \tag{14}$$

where $I_{UCA}$ is the average intensity of the UCA region and $I_{tissue}$ is that of the tissue region. The two regions were enclosed in a rectangular area (the upper white one is the tissue region and the lower red one is the UCA region).

Figure 9 shows the CTR curve between different beamformers under different transmit conditions. The performance of bubble region detection was remarkable. Figure 10 explains that, with bubble region detection, the CTR has an enhancement of 7.5 dB on average compared with ESBMV, and an enhancement of 4.2 dB compared with BWT.

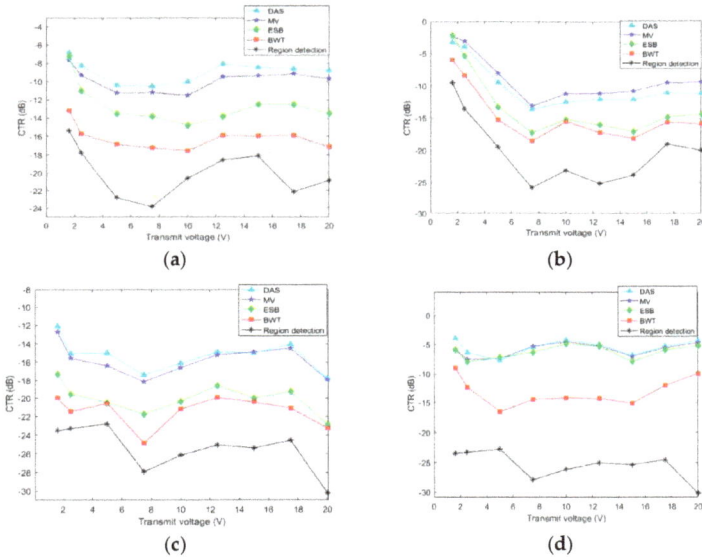

**Figure 9.** The image contrast-to-tissue ratio (CTR) curve between different beamformers: (**a**) 3 MHz; (**b**) 4 MHz; (**c**) 5 MHz; and (**d**) 6.25 MHz.

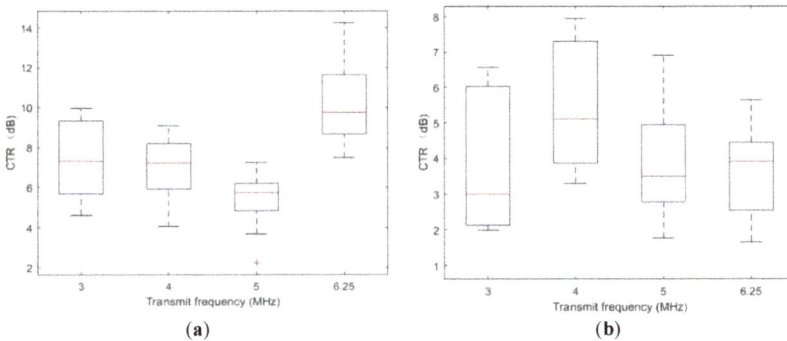

**Figure 10.** Increase in CTR with bubble region detection: (**a**) Increase in CTR between region detection and ESBMV; and (**b**) increase in CTR between region detection and BWT.

In the bubble region detection, the choice of eigenvalue and ESBCF index are two critical factors. Taking the pork experiment as an experiment, the pork and artifact disturbance remain under a low threshold while part of the UCA information is also removed with a high threshold, as shown in Figure 11.

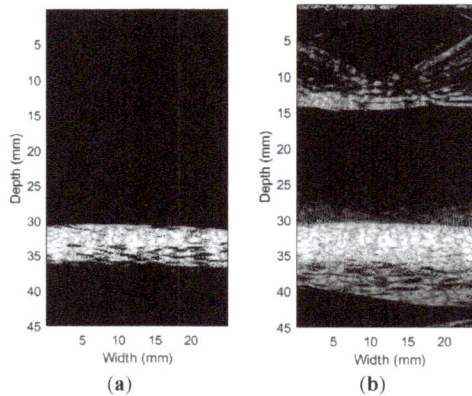

**Figure 11.** The image result of bubble region detection under different thresholds: (**a**) Higher threshold and (**b**) lower threshold.

*3.2. In Vivo Results*

To demonstrate the effectiveness of the proposed algorithm, we also performed in vivo rabbit experiments. The ear vein and kidney of rabbit were studied, respectively. Figure 12 shows a scene from the rabbit experiment targeting the kidney.

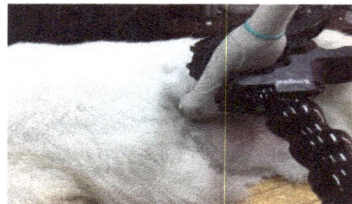

**Figure 12.** The in vivo rabbit experiment targeting the kidney.

The rabbit (2 kg, 4 months old) was first anesthetized and then placed on an autopsy table where the four limbs were fixed by ropes. Before imaging, the region of interest was epilated to remove the influence of cony hair. Medical ultrasonic coupling agent was applied to the region of interest. A total of 500 μL Sonovue microbubbles (no dilution) were injected through the right ear vein, which was followed by 500 μL of physiological saline.

The image results after microbubble injection are shown below (3 MHz, 10 V), and we added the detected UCA region to the DAS image. Figure 13 is the image targeting the ear vein ($c = 1$) and Figure 14 is the kidney ($c = 1$).

Bubble region detection can remove the noise around the kidney edge effectively, and makes the image clearer. However, the upper part of the muscle and part of the speckle noise in the image cannot be eradicated. This indicates that the tissue signal and the UCA signal still have partial overlap in terms of eigenvalues and ESBCF.

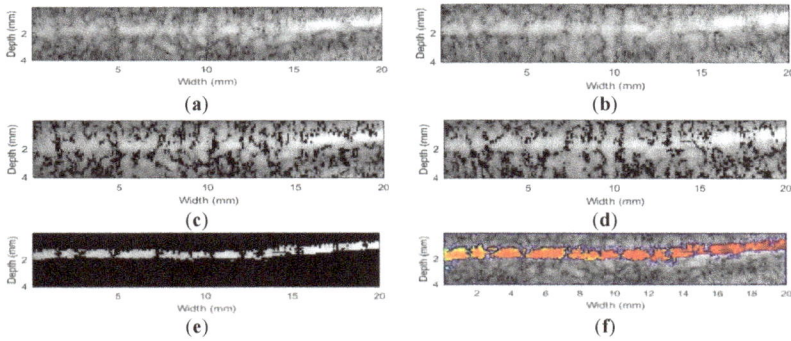

**Figure 13.** The image result of ear vein: (**a**) DAS; (**b**) MV; (**c**) ESBMV; (**d**) BWT; (**e**) bubble region detection; and (**f**) UCA-enhanced image.

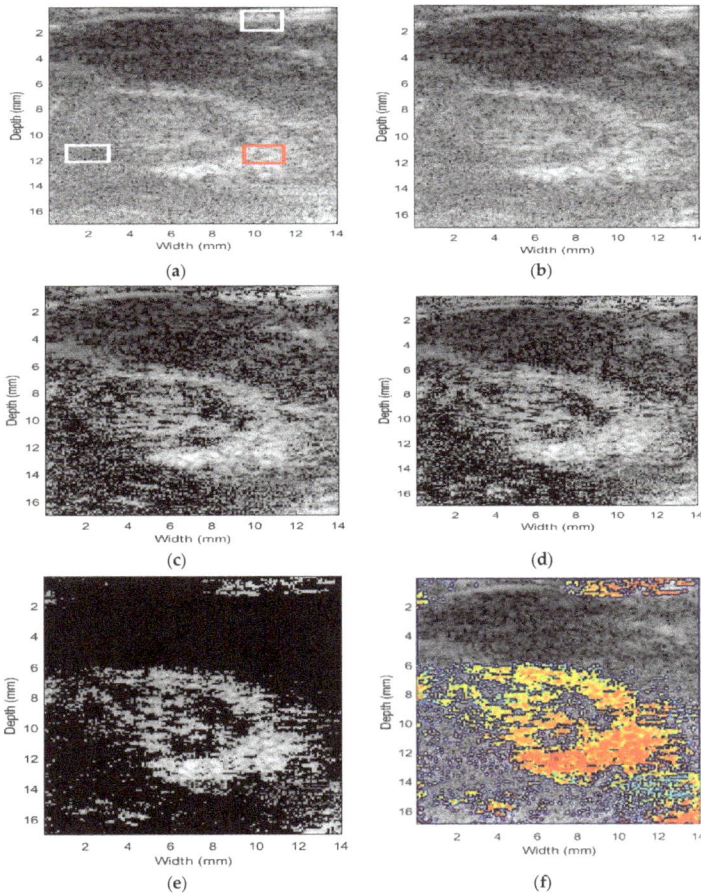

**Figure 14.** The image result of kidney: (**a**) DAS; (**b**) MV; (**c**) ESBMV; (**d**) BWT; (**e**) bubble region detection; and (**f**) UCA -enhanced image.

Taking the intensity in the red rectangle as the UCA and the white rectangles as the tissue region (CTR1 is calculated for the tissue and UCA regions with the same depth, while CTR2 is calculated with the same width, see Figure 14a), the CTR of different beamformers is shown in Table 3. CTR1 and CTR2 are improved by 18.0 dB and 18.0 dB compared with PWI based on DAS, while this improvement is 3.4 dB and 7.3 dB compared to imaging based on BWT.

**Table 3.** The image CTR of different beamformers.

| Beamformer | CTR1 (dB) | CTR2 (dB) |
|---|---|---|
| DAS | −10.1 | −4.9 |
| MV | −10.3 | −5.2 |
| ESBMV | −20.6 | −12.6 |
| BWT | −24.7 | −15.6 |
| Bubble region detection | −28.1 | −22.9 |

## 4. Conclusions

The major purpose of USCAI is to enhance the contrast between microbubble perfusion regions and surrounding regions. In this paper, a new USCAI method based on bubble region detection was proposed. This method is used to maximize the role of USCAI by taking the dissimilarity of tissue and UCA in both frequency and spatial domains into account. BWT is used to highlight UCA information from the time-frequency domain. The UCA and tissue region is further detected by utilizing their differences with the combination of maximum eigenvalue and ESBCF index. Both phantom and in vivo rabbit experiments were designed to evaluate the performance of our proposed method. It is demonstrated that the bubble edge detection we proposed provides a significant enhancement in CTR, outperforming ESBMV and BWT. The phantom and in vivo experimental results show the potential of our method for filtering out the interfering components and retaining the information of microbubbles.

**Acknowledgments:** This work was supported by the National Basic Research Program of China (61471125).

**Author Contributions:** Y.H. and Y.T. designed and carried out the experimental studies and drafted the manuscript; J.Y., as a supervisor, revised the manuscript; J.Y., Y.W., Q.Z and L.C. discussed the method and results; and S.L. participated in the design of the experiment. All authors read and approved the final manuscript.

**Conflicts of Interest:** The authors declare no conflicts of interest.

## References

1. Cosgrove, D. Ultrasound contrast agents: An overview. *Eur. J. Radiol.* **2006**, *60*, 324–330. [CrossRef] [PubMed]
2. Quaia, E. Microbubble ultrasound contrast agents: An update. *Eur. Radiol.* **2007**, *17*, 1995–2008. [CrossRef] [PubMed]
3. Tang, M.X.; Kamiyama, N.; Eckersley, R.J. Effects of Nonlinear Propagation in Ultrasound Contrast Agent Imaging. *Ultrasound Med. Biol.* **2010**, *36*, 459–466. [CrossRef] [PubMed]
4. Wilson, S.R.; Burns, P.N. Microbubble-enhanced US in body imaging: What role? *Radiology* **2010**, *257*, 24–39. [CrossRef] [PubMed]
5. Blomley, M.J.K.; Cooke, J.C.; Unger, E.C.; Monaghan, M.J.; Cosgrove, D.O. Science, Medicine, and the Future: Microbubble Contrast Agents: A New Era in Ultrasound. *BMJ Br. Med. J.* **2001**, *322*, 1222–1225. [CrossRef]
6. Couture, O.; Fink, M.; Tanter, M. Ultrasound contrast plane wave imaging. *IEEE Trans. Ultrason. Ferroelectr. Freq. Control* **2012**, *59*, 2676–2683. [CrossRef] [PubMed]
7. Tremblaydarveau, C.; Williams, R.; Milot, L.; Bruce, M.; Burns, P.N. Combined perfusion and doppler imaging using plane-wave nonlinear detection and microbubble contrast agents. *IEEE Trans. Ultrason. Ferroelectr. Freq. Control* **2014**, *61*, 1988–2000. [CrossRef] [PubMed]
8. Smeenge, M.; Mischi, M. Novel contrast-enhanced ultrasound imaging in prostate cancer. *World J. Urol.* **2011**, *29*, 581–587. [CrossRef] [PubMed]
9. Wilkening, W.; Krueger, M.; Ermert, H. Phase-coded pulse sequence for non-linear imaging. In Proceedings of the IEEE Ultrasonics Symposium (IUS), San Juan, Puerto Rico, 22–25 October 2000.

10. Wang, D.; Xuan, Y.; Wan, J.; Jing, B.; Lei, Z.; Wan, M. Ultrasound contrast plane wave imaging with higher CTR based on pulse inversion bubble wavelet transform. In Proceedings of the IEEE International Ultrasonics Symposium Ius (IUS), Chicago, IL, USA, 3–6 September 2014; pp. 1762–1765.
11. Wang, D.; Zong, Y.; Yang, X.; Hu, H.; Wan, J.; Zhang, L.; Bouakaz, A.; Wan, M. Ultrasound contrast plane wave imaging based on bubble wavelet transform: In vitro and in vivo validations. *Ultrasound Med. Biol.* **2016**, *42*, 1584–1597. [CrossRef] [PubMed]
12. Burns, P.N.; Wilson, S.R.; Simpson, D.H. Pulse inversion imaging of liver blood flow: Improved method for characterizing focal masses with microbubble contrast. *Investig. Radiol.* **2010**, *35*, 58–71. [CrossRef]
13. Eckersley, R.J.; Chin, C.T.; Burns, P.N. Optimising phase and amplitude modulation schemes for imaging microbubble contrast agents at low acoustic power. *Ultrasound Med. Biol.* **2005**, *31*, 213–219. [CrossRef] [PubMed]
14. Shen, C.C.; Lin, C.H. Chirp-encoded excitation for dual-frequency ultrasound tissue harmonic imaging. *IEEE Trans. Ultrason. Ferroelectr. Freq. Control* **2012**, *59*, 2420–2430. [PubMed]
15. Shen, C.C.; Shi, T.Y. Golay-encoded excitation for dual-frequency harmonic detection of ultrasonic contrast agents. *IEEE Trans. Ultrason. Ferroelectr. Freq. Control* **2011**, *58*, 349–356. [CrossRef] [PubMed]
16. Shekhar, H.; Doyley, M.M. The response of phospholipid-encapsulated microbubbles to chirp-coded excitation: Implications for high-frequency nonlinear imaging. *J. Acoust. Soc. Am.* **2013**, *133*, 3145–3158. [CrossRef] [PubMed]
17. Doinikov, A.A.; Bouakaz, A. Review of shell models for contrast agent microbubbles. *IEEE Trans. Ultrason. Ferroelectr. Freq. Control* **2011**, *58*, 981–993. [CrossRef] [PubMed]
18. Lu, S.; Xu, S.; Liu, R.; Hu, H.; Wan, M. High-contrast active cavitation imaging technique based on multiple bubble wavelet transform. *J. Acoust. Soc. Am.* **2016**, *140*, 1000–1011. [CrossRef] [PubMed]
19. Liu, R.; Xu, S.; Hu, H.; Huo, R.; Wang, S.; Wan, M. Wavelet-transform-based active imaging of cavitation bubbles in tissues induced by high intensity focused ultrasound. *J. Acoust. Soc. Am.* **2016**, *140*, 798–805. [CrossRef] [PubMed]
20. Doinikov, A.A.; Haac, J.F.; Dayton, P.A. Modeling of nonlinear viscous stress in encapsulating shells of lipid-coated contrast agent microbubbles. *Ultrasonics* **2009**, *49*, 269–275. [CrossRef] [PubMed]
21. Sijl, J.; Overvelde, M.; Dollet, B.; Garbin, V.; de Jong, N.; Lohse, D.; Versluis, M. "Compression-only" behavior: A second-order nonlinear response of ultrasound contrast agent microbubbles. *J. Acoust. Soc. Am.* **2011**, *129*, 1729–1739. [CrossRef] [PubMed]
22. Faez, T.; Emmer, M.; Kooiman, K.; Versluis, M.; van der Steen, A.F.; de Jong, N. 20 years of ultrasound contrast agent modeling. *IEEE Trans. Ultrason. Ferroelectr. Freq. Control* **2013**, *60*, 7–20. [CrossRef] [PubMed]
23. Capon, J. High-resolution frequency-wavenumber spectrum analysis. *Proc. IEEE* **2005**, *57*, 1408–1418. [CrossRef]
24. Asl, B.M.; Mahloojifar, A. Eigenspace-based minimum variance beamforming applied to medical ultrasound imaging. *IEEE Trans. Ultrason. Ferroelectr. Freq. Control* **2010**, *57*, 2381–2390. [CrossRef] [PubMed]
25. Johnstone, I.M. On the Distribution of the Largest Eigenvalue in Principal Components Analysis. *Ann. Stat.* **2001**, *29*, 295–327. [CrossRef]
26. Zeng, Y.; Liang, Y.C. Eigenvalue based Spectrum Sensing Algorithms for Cognitive Radio. *IEEE Trans. Commun.* **2008**, *57*, 1784–1793. [CrossRef]
27. Kortun, A.; Ratnarajah, T.; Sellathurai, M.; Liang, Y.C.; Zeng, Y. Throughput analysis using eigenvalue based spectrum sensing under noise uncertainty. In Proceedings of the Wireless Communications and Mobile Computing Conference, Limassol, Cyprus, 27–31 August 2012; pp. 395–400.
28. Zeng, Y.; Liang, Y.C.; Peh, E.C.Y.; Hoang, A.T. Cooperative covariance and eigenvalue based detections for robust sensing. In Proceedings of the Global Telecommunications Conference (GLOBECOM 2009), Honolulu, HI, USA, 30 November–4 December 2009; pp. 1–6.
29. Hollman, K.W.; Rigby, K.W.; O'Donnell, M. Coherence factor of speckle from a multi-row probe. *Proc. IEEE* **1999**, *2*, 1257–1260.
30. Xu, M.; Yang, X.; Ding, M.; Yuchi, M. Spatio-temporally smoothed coherence factor for ultrasound imaging. *IEEE Trans. Ultrason. Ferroelectr. Freq. Control* **2014**, *61*, 182–190. [CrossRef] [PubMed]
31. Wei, G.; Wang, Y.; Yu, J. Ultrasound harmonic enhanced imaging using eigenspace-based coherence factor. *Ultrasonics* **2016**, *72*, 106–116.

![applied sciences logo] *applied sciences*

MDPI

*Review*

# Riding the Plane Wave: Considerations for In Vivo Study Designs Employing High Frame Rate Ultrasound

Jason S. Au [1,2], Richard L. Hughson [2,3] and Alfred C. H. Yu [1,2,*]

[1]  Department of Electrical and Computer Engineering, University of Waterloo, Waterloo, ON N2L 3G1, Canada; jason.au@uwaterloo.ca

[2]  Schlegel-University of Waterloo Research Institute for Aging, Waterloo, ON N2J 0E2, Canada; hughson@uwaterloo.ca

[3]  Department of Kinesiology, University of Waterloo, Waterloo, ON N2L 3G1, Canada

*  Correspondence: alfred.yu@uwaterloo.ca; Tel.: +1-519-888-4567 (ext. 36908)

Received: 15 January 2018; Accepted: 13 February 2018; Published: 14 February 2018

**Abstract:** Advancements in diagnostic ultrasound have allowed for a rapid expansion of the quantity and quality of non-invasive information that clinical researchers can acquire from cardiovascular physiology. The recent emergence of high frame rate ultrasound (HiFRUS) is the next step in the quantification of complex blood flow behavior, offering angle-independent, high temporal resolution data in normal physiology and clinical cases. While there are various HiFRUS methods that have been tested and validated in simulations and in complex flow phantoms, there is a need to expand the field into more rigorous in vivo testing for clinical relevance. In this tutorial, we briefly outline the major advances in HiFRUS, and discuss practical considerations of participant preparation, experimental design, and human measurement, while also providing an example of how these frameworks can be immediately applied to in vivo research questions. The considerations put forward in this paper aim to set a realistic framework for research labs which use HiFRUS to commence the collection of human data for basic science, as well as for preliminary clinical research questions.

**Keywords:** high frame rate ultrasound; ultrafast ultrasound; human studies; neurovascular control

## 1. Introduction

Advancements in diagnostic ultrasound have allowed for a rapid expansion of the quantity and quality of non-invasive information that clinical researchers can acquire from cardiovascular physiology. As a primary application, quantification of blood flow through Doppler ultrasound is useful for the identification of early disease states or diagnosis of pathological conditions either as consequence, or root cause, of altered hemodynamics [1,2]. These applications range across multiple organ systems, for instance, grading conduit artery stenoses by flow jet velocity [3,4], estimating mitral valve inflow for left ventricular diastolic dysfunction [5], or identifying locations at risk for atherosclerotic plaque development by low/oscillatory wall shear stress [6]. However, much of the current technology is limited by assumptions of laminar flow and slow-moving blood flow in a small acoustic window, which restrict most applications to imaging with fixed insonation angles and simple flow patterns.

In recent years, high frame rate ultrasound (HiFRUS; also termed 'ultrafast' ultrasound) imaging techniques have been developed to address the above limitations by insonating a large area with unfocused beams through either spherical or plane wave emissions [7]. Rather than being limited by line-by-line pulse-echo imaging, HiFRUS techniques acquire full-field data at very high frame rates (e.g., 1000–10,000 frames per second) for excellent temporal resolution, and allow beamforming in any direction for angle-independent blood velocity estimations. These advancements enable accurate quantification of high-velocity non-laminar flows, demonstrated both in vivo and in geometrically-realistic phantoms at vessel bifurcations [8–10], aneurysms [11], and even 3D structures such as the left ventricle [12,13]. However, the ability for basic science or clinical researchers to access the vast amount of biological information has so far been limited, with recent HiFRUS reviews touting only the perceived potential of these advances in imaging [7,14–16].

The major challenges for mainstream adoption of HiFRUS technology stem from the lack of large sample human data, as well as descriptive studies of normal and pathological physiology. In this review, we provide a practical framework for basic science researchers conducting preliminary in vivo studies with the long-term objective of clinical relevance. Here, we propose a guideline for standards in technical reporting, and provide considerations for investigations in humans, data management, and storage, as well as a narrative example of how studies could be designed for in vivo observations in basic science questions. Although this review focuses on relevant cardiovascular applications, we encourage the readers to extend these principles to other areas of research which may benefit from HiFRUS imaging.

## 2. A Synopsis of High Frame Rate Ultrasound Technology

Before we proceed to discuss how in vivo investigation protocols for HiFRUS can be designed, let us first briefly review the current state-of-the-art in HiFRUS technology. While the notion of HiFRUS imaging with sub-millisecond time resolution has been in conception since the 1980s [17], the technology has garnered significant attention since the turn of this decade [18]. Such a rapid surge of interest is technically attributed to two engineering innovation trends. First, in the past decade, reconfigurable ultrasound scanners have become more prevalent [19–24], as opposed to non-programmable clinical systems that are designed via an embedded system approach [25]. These open-architecture systems have enabled researchers to readily implement different variants of unfocused pulsing sequences that are essential for realizing HiFRUS [26]. Second, high-throughput computing hardware such as graphical processing units have greatly matured [27]. These parallel processing devices have served well to achieve real-time execution of HiFRUS-related computation tasks, such as pixel-by-pixel beamforming [28–30], Doppler processing [31–33], and post hoc filtering [34].

At present, a number of academic labs have developed in-house research scanners with HiFRUS capabilities. Worth particular mention are the in-house systems built at the Technical University of Denmark [19,23], the University of Florence [21,24], the Langevin Institute in Paris [35], and the Polish Academy of Sciences [36]. A few commercially available research platforms also allow similar HiFRUS implementations, such as Analogic Ultrasound (Peabody, MA, USA) [37], Verasonics (Kirkland, WA, USA) [22], US4US (Warsaw, Poland), and Cephasonics (Santa Clara, CA, USA). Note that most conventional ultrasound scanners cannot be readily reconfigured to implement HiFRUS because their architecture is typically developed through an embedded system design approach that only specializes in performing beamline-based imaging [25]; this is why research scanners have become essential for advances in diagnostic ultrasound. Another point worth noting is that one clinical scanner developer (Supersonic Imagine, Aux-du-Province, France) is currently dedicated to the development of the HiFRUS market [38]. Also, specialized HiFRUS flow vector imaging modes are available on clinical scanners developed by Analogic Ultrasound (Peabody, MA, USA) [39] and Mindray (Shenzhen, China) [40].

In terms of its technical principles, HiFRUS imaging is fundamentally instituted upon the pulse-echo sensing paradigm, similar to conventional ultrasound imaging. However, instead of using focused beams for transmission, HiFRUS instead uses unfocused pulsing strategies in the forms of spherical waves [41] or plane waves [18,38], as shown in Figure 1. On reception, the pulse echoes returned from the imaging plane of interest are recorded on every array transducer channel. Pixel-by-pixel beamforming is then performed using the channel-domain pulse-echo data, in which each image pixel value is derived through a "delay and sum" approach [28]. One salient point to be noted is that the lateral spatial resolution of HiFRUS is inherently not as fine as conventional ultrasound because unfocused transmit firings are used. Nonetheless, the temporal resolution is significantly improved because, from each transmit event's channel-domain data set, it is possible to form one image frame based on pulse-echoes returned from the entire imaging view. To improve the lateral resolution of HiFRUS images, one strategy that can be leveraged is to perform coherent compounding of low-resolution image frames derived from different spherical firing positions [41] or different plane wave steering angles [42]. While this compounding operation would unavoidably reduce the frame rate, it may be carried out recursively to limit the loss in frame rate [43].

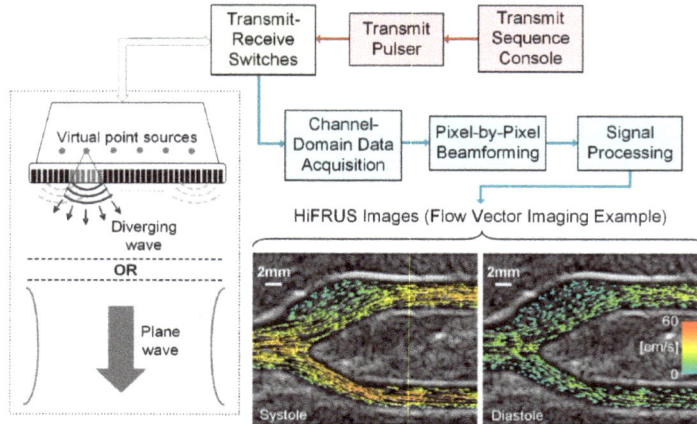

**Figure 1.** Overview of high frame rate ultrasound (HiFRUS) data acquisition principles and the signal processing chain. Two sample HiFRUS images in the form of flow vector imaging with dynamic visualization [8] are shown in the context of a carotid bifurcation model.

From an application standpoint, the key diagnostic value offered by HiFRUS is the full-field, high-resolution spatiotemporal imaging that can yield functional insight into physiological events taking place inside the human body. For instance, by integrating HiFRUS with Doppler estimation principles, it is possible to achieve time-resolved visualization of complex flow dynamics through the rendering of flow speckles [9] and the derivation of flow vectors at different pixel positions [44] as illustrated in the sample images in Figure 1. Not only is this useful in examining carotid hemodynamics [8,40], it is also applicable to the evaluation of arterial strain [45–47], the visualization of pulse wave propagation through the artery wall [48], and the tracking of shear waves propagating in tissues [35,49]. HiFRUS may also be used in urology applications to gain time-resolved insight into turbulent urinary flow behavior [50]. This technology may be used in cardiac applications to study myocardial contraction [51,52] and intraventricular flow patterns [53,54], although these techniques need further refinement. Emerging developments in HiFRUS methodologies, including the incorporation of contrast agents [55] and state-of-the-art 4D imaging [56], will undoubtedly lead to further physiological discoveries to enhance application prospects in this field.

## 3. Framework for In Vivo Cardiovascular Studies

The refinement of HiFRUS and blood flow vector quantification has led to a large number of technical descriptions and validation studies, but very few in vivo clinical studies in which the potential for advanced imaging methods can be highlighted. In these limited human studies, the focus has been on proof-of-concept study designs, with small sample sizes, limited a priori hypotheses, and case studies, rather than group comparisons or interventional designs [10,12,13,39,57–59]. Focused studies on basic science or clinical research questions are the natural next step for the field, in which

HiFRUS can be used as a specialized research tool for both simple and complex system cardiovascular measurement. However, in order to build in a level of consistency between the varied methods, HiFRUS research groups should be aware of the technical, physiological, and ethical considerations required for larger scale human studies. Below, we outline such considerations, of which we encourage for high-quality reporting and study design.

### 3.1. Validation of Methods Prior to In Vivo Data Collection

The range in both hardware and software solutions for HiFRUS platforms introduces a degree of uncertainty in the accuracy and validity of methods during the early stages of system development. As measurement error is an important component of interventional in vivo studies, the accuracy in flow imaging should be determined through validation studies, using either criterion standard methodology (e.g., magnetic resonance imaging [60]), or validated models with known hemodynamic properties (e.g., computational fluid dynamics simulations [61] and flow phantoms [62]). The specific decisions for constructing and validating flow phantoms have been previously reviewed [63], and should be considered prior to diverting resources to human studies.

### 3.2. Standards in Technical Reporting

While few clinical ultrasound systems have the capability for HiFRUS imaging, certain research systems allow for customization of the transmit firing sequence of every array channel and the acquisition of channel-domain data for customized algorithmic processing [19–24,35–37]. In general, HiFRUS imaging studies rigorously report the technical specifics of the experimental methods, such as the array pitch, probe frequency, transmit pulse duration, pulse repetition frequency, steering angles, and spherical source positioning. As HiFRUS technology is adopted into basic science and clinical research, such details should be preserved and summarized in study reports for easy comparison between methods. Table 1 lists the essential technical reporting that should be ideally described in such studies. In addition to the technical reporting, scanning locations should be rigorously reported, including the organ of interest, relevant landmarks for reproducibility, details on insonation angles (e.g., anterior or lateral plane on the neck), and target organ orientations (e.g., visualization of both the internal and external carotid arteries in the same plane).

**Table 1.** Technical reporting in high frame rate ultrasound (HiFRUS) investigations and proposed details for the research example presented in Section 4.

| HiFRUS Parameter | Value |
| --- | --- |
| Scanning system | SonixTouch |
| Array pitch | 0.3048 mm |
| Probe frequency | 5 MHz |
| Emission method | Plane wave excitation |
| Transmit pulse duration | 2 cycles |
| Pulse repetition frequency | 10 kHz |
| Steering angles | $-10°, 0°, +10°$ |
| Slow-time window size (or ensemble length) | 128 samples (12.8 ms) |
| Slow-time window step size | 4 samples (0.4 ms) |
| Effective frame rate | 833 fps |
| Scanning location | Left of image aligned 1 cm proximal to the carotid bifurcation |
| Collection duration | 3 s (16 GB on-board memory) |

The majority of clinical ultrasound studies are performed by highly trained sonographers, often hired by clinical research teams. Although research sonographers are highly trained, considerations should be given to any motion artefact caused by human error that may influence data quality during HiFRUS acquisitions. Considering the limited data that can be acquired during study sessions (potentially within a few heart cycles due to data storage and processing as discussed below), it is important to eliminate sources of variability beyond that of the biological system being investigated.

For this reason, probe holders should be considered as part of the experimental set-up when possible, and reported in the study methodology. Stereotactic probe holders have previously been shown to reduce the typical error of the flow-mediated dilation technique, which is based on diameter changes in the brachial artery after a brief period of distal limb ischemia [64].

### 3.3. Human Considerations

A shift in focus from flow phantoms to human participants brings about certain considerations for in vivo testing that may introduce unwanted variability into data quality. The human cardiovascular system is tightly regulated by the autonomic nervous system, involving beat-to-beat neurovascular regulation through the sympathetic and parasympathetic nervous systems [65,66]. However, this regulation may offer unwanted sources of variability during HiFRUS studies, as only a few cardiac cycles can realistically be acquired per participant due to the high frame rate. Decisions on participant pre-visit instructions and study methodology should be made with consideration on how to minimize these confounders. For example, technical guidelines for evaluations of carotid–femoral pulse wave velocity and flow-mediated dilation have recommended that: participants refrain from food (~2–6 h), caffeine, smoking, and alcohol (~12 h) prior to measurement; testing should occur in a quiet, temperature-controlled room after 10 min of supine rest; and measurements be taken at the same time of day for longitudinal studies to account for circadian rhythm [67–74]. While Table 2 lists recommendations for the average participant, any perturbations from the 'ambulatory' state of an individual should be considered; for example, asking an individual who habitually smokes to refrain from smoking may cause an equal amount of distress and undue sympathetic activation. As the novelty of HiFRUS data limits the available literature on the confounders of measurement variability, we suggest that the above basic study controls be considered for studies involving human participants, with the aim of reducing unwanted variability in the high-quality data.

**Table 2.** Methodological considerations in cardiovascular human testing.

| Recommendation | Reason |
| --- | --- |
| Pre-visit instructions | |
| 2 h fasted | Altered sympathetic activation |
| 6 h refrain from caffeine | Altered sympathetic activation |
| 12 h refrain from smoking | Acute effects on vascular structure and function |
| 12 h refrain from moderate-to-vigorous physical activity | Acute effects on vascular structure and function |
| 12 h refrain from alcohol | Acute effects on vascular structure and function |
| Record of current medications | Various acute and chronic effects on the vasculature |
| Participant preparation | |
| Assign unlinked participant ID | Ethical considerations for sensitive health information |
| 10 min rest period | Altered sympathetic activation upon arrival to the lab |
| Resting heart rate recording | Detail of the hemodynamic environment |
| Resting blood pressure recording | Detail of the hemodynamic environment |
| Probe holder placement | Reduction of motion artifacts |
| Breath hold during acquisition | Reduction of motion artifacts |

Medications are an important consideration when preparing to interpret the potentially clinically relevant information provided in HiFRUS examinations. As it is unethical and unsafe to ask participants to withhold all medication for certain studies, it is important to record and consider possible confounders when interpreting findings. Particular care should be taken with commonly prescribed medications for hypertension and heart disease, such as statins, beta-blockers, angiotensin-converting enzyme inhibitors, and diuretics, all of which result in measurable effects on cardiovascular function including reductions in heart rate, blood volume and arterial stiffness [67].

Further to the issue of beat-to-beat variability, additional considerations should be given to factors that may affect the ability of individual heart cycles to reflect 'steady state' conditions of an individual or physiological state. Respiratory sinus arrhythmia is a known phenomenon in physiological monitoring, eliciting predictable fluctuations in heart rate and blood pressure [75]. Given that typical respiratory and heart rates in healthy adults are ~20 breaths/min and ~70 beats/min, respectively, an average of at least six heart cycles is necessary to accurately reflect steady state physiology, which may not be feasible for all HiFRUS acquisition systems. Borrowing from echocardiography guidelines, an alternative solution to averaging is implementing brief breath holds to limit both the sympathetic and mechanical effects of breathing [76,77]. Other arrhythmias such as ectopic beats, premature atrial/ventricular contractions, or flutters may cause difficulty in recording steady state data, although arrhythmia physiology may be an interesting study area in itself for HiFRUS examination.

Variable acoustic windows and poor image quality are potential barriers for high quality HiFRUS investigations, which may limit the available participant pool in human studies. While flow phantoms offer complete control of model orientation, depth, and acoustic medium, human studies will certainly produce sub-optimal data quality, which will vary between methods. An additional concern for data quality is the potential waste of time and resources that would accompany data dropout, especially for repeated-measures study designs. To address this concern, we recommend including image quality as an exclusion criterion in ethics applications, whereby individuals are first consented under the local ethics board and are then screened for image quality before the start of the study protocol. If participants are being remunerated for their time, they would have to be provided a pro-rated amount as they have been officially enrolled in the protocol upon giving consent. Regardless of the participant flow, reporting quality metrics, such as the number of unusable images or number and reason for data loss, will provide valuable transparency on HiFRUS methodology. Although designed for randomized control trial use, the CONSORT guidelines provide excellent descriptions of high quality participant reporting, which includes number of participants excluded, participants lost to follow up in repeated measures designs, and participants excluded from analysis [78].

## 4. Research Example: Neurovascular Control and Complex Blood Flow

The above general recommendations for in vivo investigations provide a general framework under which the acquisition and management of human data should be optimally performed in cardiovascular research. In order to supplement these considerations, below we present an example of a simple and realistic study design for a basic science research question, describing each step in the design process from research question to data acquisition.

The most attractive element of HiFRUS for a cardiovascular physiologist is the unprecedented quantity and quality of information that can non-invasively be assessed from the conduit arteries. High temporal and spatial sensitivity of complex blood flow patterns may offer a unique view of traditional cardiovascular techniques that have otherwise been well established in the field. For example, the cold pressor test (CPT) is a well-documented assessment of neurovascular reactivity [79,80], which has a history of central and peripheral responder sub-types [81], and documented effects on intra- and extra-cranial blood flow [82,83]. The CPT is highlighted by its simple protocol: submersion of either the hand or foot into an ice bath (~2–4 °C) for a duration of ~2 min. This stimulus elicits a rapid neurovascular response, characterized by $\alpha$-adrenergic peripheral vasoconstriction causing slight increases in heart rate (i.e., +5 to +10 beats per minute) and moderate increases in mean arterial pressure (i.e., +15 to +25 mmHg). While recent studies have reported increases in common carotid, internal carotid, and middle cerebral artery blood velocity with the CPT [82,83], it may be of value to characterize the complex flow patterns at the carotid bifurcation to further detail the known hemodynamics of the CPT, as well as to perhaps develop an easily accessible reactivity test for carotid bifurcation jets and recirculation zones under challenge conditions. From this knowledge gap in a well-established cardiovascular technique, we can design an acute interventional within-subject human research question that directly uses the novel capabilities of HiFRUS with a priori hypotheses: does the CPT elicit changes in carotid bifurcation flow jet velocity or recirculation zones as measured by HiFRUS?

Figure 2 and Table 1 detail the methodological and technical design elements that we would employ for such a study. Young healthy adults are the ideal participants for this design, as we would like to test a basic science question in a controlled system, ideally by manipulating only the variables of interest without considering confounding pathology. After institutional safety and ethics approval, participants would be recruited and consented for testing in the lab. To ensure quality data, we suggest that potential participants be screened for appropriate scanning windows for appropriate orientation of the carotid bifurcation (i.e., at minimum, in-plane visualization of the carotid bulb and internal carotid artery). This exclusion criteria limits the generalizability of the experimental findings, but we must acknowledge the variability in carotid geometry in the general population using 2D ultrasound [84,85]. To limit confounding factors for measurement variability, we would ask participants to arrive at the lab having fasted for two hours, and having refrained from moderate-to-vigorous physical activity, smoking, and alcohol in the 12 h prior to assessments (Table 2).

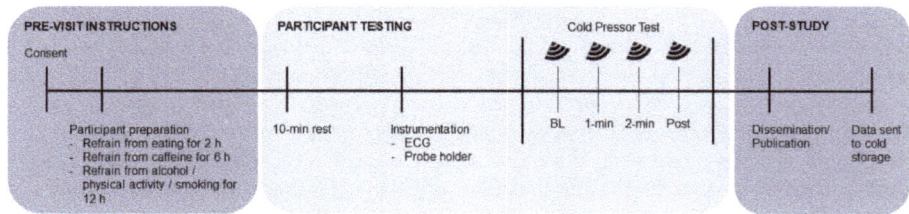

**Figure 2.** An example of a study timeline and protocol for the experiment outlined in the Research Example. ECG: electrocardiogram.

During the testing protocol, participants would rest supine for at least ten minutes prior to data collection to standardize the hemodynamic environment (i.e., heart rate and blood pressure). For this particular research question, it would be valuable to gate analysis to the electrocardiogram (ECG) trace, in order to account for some beat-to-beat variability in blood flow, as well as to assist with aligning secondary data acquisition such as beat-to-beat blood pressure finger plethysmography or transcranial Doppler signals. A simple single-lead ECG can be used to align the data either on the ultrasound unit, or through simultaneous capture with an external data acquisition system. If possible, participants would be instrumented with a stereotactic probe holder to standardize the anterior–posterior orientation of the ultrasound probe to assist with repeated assessments in the same plane. During acquisition itself, participants would be asked to briefly hold their breath while six heart cycles (~6–8 s; ~16 GB on our system) are recorded. The specific details of the HiFRUS acquisition and analysis would vary between research groups, although we recommend the technical reporting suggested in Table 1 for a fully detailed methods sections to be included for publication.

The above research example is just one of many avenues that HiFRUS methodology may take in cardiovascular science (for other clinical examples, see [7,14–16]). Diagnostic testing in clinical studies presents a unique set of challenges (e.g., power calculations for novel outcomes [86], specificity and sensitivity to detect abnormal hemodynamics [87]), the discussion of which is beyond the scope of this review. Although individual research questions, experimental protocols, and outcome measures may vary, we hope this general description of a research outline may prove useful to researchers beginning to explore the potential for HiFRUS investigations in human participants.

## 5. Summary

HiFRUS techniques for complex blood flow quantification are being rapidly developed, but have yet to be implemented in larger scale human studies investigating either basic science or clinical research questions. In this paper, we have put forward methodological and technical reporting aspects that should be considered as part of future study designs in the area. Participant preparation, variability controls in experimental protocols, and ethical considerations are just a few of the points that will elevate the research standards in HiFRUS investigations, which we highlight in a practical example in neurovascular control of blood flow. As these techniques are eventually adopted into larger scale studies, we encourage ultrasound researchers to reach out to partner with physiologists and clinical researchers and extend the possibilities for HiFRUS as an invaluable research tool in cardiovascular science.

**Acknowledgments:** J.S.A. is funded by a Canadian Institutes of Health Research (CIHR) Post-doctoral Fellowship (MFE-152454). Research funding from the CIHR Project Grant (PJT-153240), Natural Sciences and Engineering Research Council of Canada (RGPIN-2016-04042), Canada Foundation for Innovation (36138), and the Ontario Early Researcher Award (ER16-12-186) is also gratefully acknowledged.

**Author Contributions:** J.S.A. and A.C.H.Y. contributed to the drafting of this manuscript. J.S.A., R.L.H. and A.C.H.Y. edited and approved the final version of the manuscript.

**Conflicts of Interest:** The authors declare no conflict of interest.

## References

1. Quiñones, M.A.; Otto, C.M.; Stoddard, M.; Waggoner, A.; Zoghbi, W.A. Recommendations for quantification of Doppler echocardiography: A report from the Doppler quantification task force of the nomenclature and standards committee of the American Society of Echocardiography. *J. Am. Soc. Echocardiogr.* **2002**, *15*, 167–184. [CrossRef] [PubMed]

2. Rooke, T.W.; Hirsch, A.T.; Misra, S.; Sidawy, A.N.; Beckman, J.A.; Findeiss, L.K.; Golzarian, J.; Gornik, H.L.; Halperin, J.L.; Jaff, M.R.; et al. 2011 ACCF/AHA focused update of the guideline for the management of patients with peripheral artery disease (Updating the 2005 guideline). *Catheter. Cardiovasc. Interv.* **2012**, *79*, 501–531. [CrossRef] [PubMed]

3. Steel, R.; Ramnarine, K.V.; Davidson, F.; Fish, P.J.; Hoskins, P.R. Angle-independent estimation of maximum velocity through stenoses using vector Doppler ultrasound. *Ultrasound Med. Biol.* **2003**, *29*, 575–584. [CrossRef]

4. Oates, C.P.; Naylor, A.R.; Hartshorne, T.; Charles, S.M.; Fail, T.; Humphries, K.; Aslam, M.; Khodabakhsh, P. Joint recommendations for reporting carotid ultrasound investigations in the United Kingdom. *Eur. J. Vasc. Endovasc. Surg.* **2009**, *37*, 251–261. [CrossRef] [PubMed]

5. Nagueh, S.F.; Smiseth, O.A.; Appleton, C.P.; Byrd, B.F.; Dokainish, H.; Edvardsen, T.; Flachskampf, F.A.; Gillebert, T.C.; Klein, A.L.; Lancellotti, P.; et al. Recommendations for the evaluation of left ventricular diastolic function by echocardiography: An update from the American Society of Echocardiography and the European Association of Cardiovascular Imaging. *Eur. Heart J. Cardiovasc. Imaging* **2016**, *17*, 1321–1360. [CrossRef] [PubMed]

6. Cheng, C.; Tempel, D.; Van Haperen, R.; Van Der Baan, A.; Grosveld, F.; Daemen, M.J.A.P.; Krams, R.; De Crom, R. Atherosclerotic lesion size and vulnerability are determined by patterns of fluid shear stress. *Circulation* **2006**, *113*, 2744–2753. [CrossRef] [PubMed]

7. Jensen, J.A.; Nikolov, S.; Yu, A.C.H.; Garcia, D. Ultrasound vector flow imaging—Part II: Parallel systems. *IEEE Trans. Ultrason. Ferroelectr. Freq. Control* **2016**, *63*, 1722–1732. [CrossRef] [PubMed]

8. Yiu, B.Y.S.; Lai, S.S.M.; Yu, A.C.H. Vector projectile imaging: Time-resolved dynamic visualization of complex flow patterns. *Ultrasound Med. Biol.* **2014**, *40*, 2295–2309. [CrossRef] [PubMed]

9. Yiu, B.Y.S.; Yu, A.C.H. High-frame-rate ultrasound color-encoded speckle imaging of complex flow dynamics. *Ultrasound Med. Biol.* **2013**, *39*, 1015–1025. [CrossRef] [PubMed]

10. Hansen, P.M.; Pedersen, M.M.; Hansen, K.L.; Nielsen, M.B.; Jensen, J.A. Demonstration of a vector velocity technique. *Ultraschall Med.* **2011**, *32*, 213–215. [CrossRef] [PubMed]

11. Ho, C.K.; Chee, A.J.Y.; Yiu, B.Y.S.; Tsang, A.C.O.; Chow, K.W.; Yu, A.C.H. Wall-less flow phantoms with tortuous vascular geometries: Design principles and a patient-specific model fabrication example. *IEEE Trans. Ultrason. Ferroelectr. Freq. Control* **2017**, *64*, 25–38. [CrossRef] [PubMed]

12. Garcia, D.; del Alamo, J.C.; Tanne, D.; Yotti, R.; Cortina, C.; Bertrand, E.; Antoranz, J.C.; Perez-David, E.; Rieu, R.; Fernandez-Aviles, F.; et al. Two-dimensional intraventricular flow mapping by digital processing conventional color-doppler echocardiography images. *IEEE Trans. Med. Imaging* **2010**, *29*, 1701–1713. [CrossRef] [PubMed]

13. Hendabadi, S.; Bermejo, J.; Benito, Y.; Yotti, R.; Fernández-Avilés, F.; Del Álamo, J.C.; Shadden, S.C. Topology of blood transport in the human left ventricle by novel processing of doppler echocardiography. *Ann. Biomed. Eng.* **2013**, *41*, 2603–2616. [CrossRef] [PubMed]

14. Jensen, J.A.; Nikolov, S.I.; Yu, A.C.H.; Garcia, D. Ultrasound vector flow imaging—Part I: Sequential systems. *IEEE Trans. Ultrason. Ferroelectr. Freq. Control* **2016**, *63*, 1704–1721. [CrossRef] [PubMed]

15. Sengupta, P.P.; Pedrizzetti, G.; Kilner, P.J.; Kheradvar, A.; Ebbers, T.; Tonti, G.; Fraser, A.G.; Narula, J. Emerging trends in CV flow visualization. *JACC Cardiovasc. Imaging* **2012**, *5*, 305–316. [CrossRef] [PubMed]

16. Goddi, A.; Fanizza, M.; Bortolotto, C.; Raciti, M.V.; Fiorina, I.; He, X.; Du, Y.; Calliada, F. Vector flow imaging techniques: An innovative ultrasonographic technique for the study of blood flow. *J. Clin. Ultrasound* **2017**, *45*, 582–588. [CrossRef] [PubMed]

17. Shattuck, D.P.; Weinshenker, M.D.; Smith, S.W.; von Ramm, O.T. Explososcan: A parallel processing technique for high speed ultrasound imaging with linear phased arrays. *J. Acoust. Soc. Am.* **1984**, *75*, 1273–1282. [CrossRef] [PubMed]

18. Tanter, M.; Fink, M. Ultrafast imaging in biomedical ultrasound. *IEEE Trans. Ultrason. Ferroelectr. Freq. Control* **2014**, *61*, 102–119. [CrossRef] [PubMed]

19. Jensen, J.A.; Holm, O.; Jensen, L.J.; Bendsen, H.; Nikolov, S.I.; Tomov, B.G.; Munk, P.; Hansen, M.; Salomonsen, K.; Hansen, J.; et al. Ultrasound research scanner for real-time synthetic aperture data acquisition. *IEEE Trans. Ultrason. Ferroelectr. Freq. Control* **2005**, *52*, 881–891. [CrossRef] [PubMed]

20. Lu, J.Y.; Cheng, J.; Wang, J. High frame rate imaging system for limited diffraction array beam imaging with square-wave aperture weightings. *IEEE Trans. Ultrason. Ferroelectr. Freq. Control* **2006**, *53*, 1796–1811. [CrossRef] [PubMed]

21. Tortoli, P.; Bassi, L.; Boni, E.; Dallai, A.; Guidi, F.; Ricci, S. ULA-OP: An advanced open platform for ultrasound research. *IEEE Trans. Ultrason. Ferroelectr. Freq. Control* **2009**, *56*, 2207–2216. [CrossRef] [PubMed]

22. Daigle, R.E. Ultrasound Imaging System with Pixel Oriented Processing. U.S. Patent 8,287,456B2, 16 October 2012.

23. Jensen, J.A.; Holten-Lund, H.; Nilsson, R.T.; Hansen, M.; Larsen, U.D.; Domsten, R.P.; Tomov, B.G.; Stuart, M.B.; Nikolov, S.I.; Pihl, M.J.; et al. SARUS: A synthetic aperture real-time ultrasound system. *IEEE Trans. Ultrason. Ferroelectr. Freq. Control* **2013**, *60*, 1838–1852. [CrossRef] [PubMed]

24. Boni, E.; Bassi, L.; Dallai, A.; Guidi, F.; Meacci, V.; Ramalli, A.; Ricci, S.; Tortoli, P. ULA-OP 256: A 256-channel open scanner for development and real-time implementation of new ultrasound methods. *IEEE Trans. Ultrason. Ferroelectr. Freq. Control* **2016**, *63*, 1488–1495. [CrossRef] [PubMed]

25. Powers, J.; Kremkau, F. Medical ultrasound systems. *Interface Focus* **2011**, *1*, 477–489. [CrossRef] [PubMed]

26. Boni, E.; Bassi, L.; Dallai, A.; Meacci, V.; Ramalli, A.; Scaringella, M.; Guidi, F.; Ricci, S.; Tortoli, P. Architecture of an ultrasound system for continuous real-time high frame rate imaging. *IEEE Trans. Ultrason. Ferroelectr. Freq. Control* **2017**, *64*, 1276–1284. [CrossRef] [PubMed]

27. So, H.; Chen, J.; Yiu, B.; Yu, A. Medical ultrasound imaging: To GPU or not to GPU? *IEEE Micro* **2011**, *31*, 54–65. [CrossRef]

28. Yiu, B.Y.S.; Tsang, I.K.H.; Yu, A.C.H. GPU-based beamformer: Fast realization of plane wave compounding and synthetic aperture imaging. *IEEE Trans. Ultrason. Ferroelectr. Freq. Control* **2011**, *58*, 1698–1705. [CrossRef] [PubMed]

29. Martin-Arguedas, C.J.; Romero-Laorden, D.; Martinez-Graullera, O.; Perez-Lopez, M.; Gomez-Ullate, L. An ultrasonic imaging system based on a new SAFT approach and a GPU beamformer. *IEEE Trans. Ultrason. Ferroelectr. Freq. Control* **2012**, *59*, 1402–1412. [CrossRef] [PubMed]

30. Yiu, B.Y.S.; Yu, A.C.H. GPU-based minimum variance beamformer for synthetic aperture imaging of the eye. *Ultrasound Med. Biol.* **2015**, *41*, 871–883. [CrossRef] [PubMed]

31. Chang, L.W.; Hsu, K.H.; Li, P.C. Graphics processing unit-based high-frame-rate color doppler ultrasound processing. *IEEE Trans. Ultrason. Ferroelectr. Freq. Control* **2009**, *56*, 1856–1860. [CrossRef] [PubMed]

32. Rosenzweig, S.; Palmeri, M.; Nightingale, K. GPU-based real-time small displacement estimation with ultrasound. *IEEE Trans. Ultrason. Ferroelectr. Freq. Control* **2011**, *58*, 399–405. [CrossRef] [PubMed]

33. Chee, A.J.Y.; Yiu, B.Y.S.; Yu, A.C.H. A GPU-parallelized Eigen-based clutter filter framework for ultrasound color flow imaging. *IEEE Trans. Ultrason. Ferroelectr. Freq. Control* **2017**, *64*, 150–163. [CrossRef] [PubMed]

34. Broxvall, M.; Emilsson, K.; Thunberg, P. Fast GPU based adaptive filtering of 4D echocardiography. *IEEE Trans. Med. Imaging* **2012**, *31*, 1165–1172. [CrossRef] [PubMed]

35. Tanter, M.; Bercoff, J.; Sandrin, L.; Fink. M. Ultrafast compound imaging for 2-D motion vector estimation: Application to transient elastography. *IEEE Trans. Ultrason. Ferroelectr. Freq. Control* **2002**, *49*, 1363–1374. [CrossRef] [PubMed]

36. Lewandowski, M.; Walczak, M.; Witek, B.; Kulesza, P.; Sielewicz, K. Modular & scalable ultrasound platform with GPU processing. In Proceedings of the 2012 IEEE International Ultrasonics Symposium (IUS), Dresden, Germany, 7–10 October 2012. [CrossRef]

37. Cheung, C.; Yu, A.; Salimi, N.; Yiu, B.; Tsang, I.; Kerby, B.; Azar, R.; Dickie, K. Multi-channel pre-beamformed data acquisition system for research on advanced ultrasound imaging methods. *IEEE Trans. Ultrason. Ferroelectr. Freq. Control* **2012**, *59*, 243–253. [CrossRef] [PubMed]

38. Bercoff, J. Ultrafast ultrasound imaging. In *Ultrasound Imaging-Medical Applications*; Minin, I.V., Minin, O.V., Eds.; InTech: New York, NY, USA, 2011; pp. 3–24.

39. Hansen, K.L.; Møller-Sørensen, H.; Pedersen, M.M.; Hansen, P.M.; Kjaergaard, J.; Lund, J.T.; Nilsson, J.C.; Jensen, J.A.; Nielsen, M.B. First report on intraoperative vector flow imaging of the heart among patients with healthy and diseased aortic valves. *Ultrasonics* **2015**, *56*, 243–250. [CrossRef] [PubMed]

40. Goddi, A.; Bortolotto, C.; Fiorina, I.; Raciti, M.V.; Fanizza, M.; Turpini, E.; Boffelli, G.; Calliada, F. High-frame rate vector flow imaging of the carotid bifurcation. *Insights Imaging* **2017**, *8*, 319–328. [CrossRef] [PubMed]

41. Jensen, J.A.; Nikolov, S.I.; Gammelmark, K.L.; Pedersen, M.H. Synthetic aperture ultrasound imaging. *Ultrasonics* **2006**, *44*, e5–e15. [CrossRef] [PubMed]

42. Montaldo, G.; Tanter, M.; Bercoff, J.; Benech, N.; Fink, M. Coherent plane-wave compounding for very high frame rate ultrasonography and transient elastography. *IEEE Trans. Ultrason. Ferroelectr. Freq. Control* **2009**, *56*, 489–506. [CrossRef] [PubMed]

43. Nikolov, S.I.; Jensen, J.A. In-vivo synthetic aperture flow imaging in medical ultrasound. *IEEE Trans. Ultrason. Ferroelectr. Freq. Control* **2003**, *50*, 848–856. [CrossRef] [PubMed]

44. Yiu, B.Y.S.; Yu, A.C.H. Least-squares multi-angle Doppler estimators for plane-wave vector flow imaging. *IEEE Trans. Ultrason. Ferroelectr. Freq. Control* **2016**, *63*, 1733–1744. [CrossRef] [PubMed]

45. Hansen, H.H.G.; Saris, A.E.C.M.; Vaka, N.R.; Nillesen, M.M.; de Korte, C.L. Ultrafast vascular strain compounding using plane wave transmission. *J. Biomech.* **2014**, *47*, 815–823. [CrossRef] [PubMed]

46. Kruizinga, P.; Mastik, F.; van den Oord, S.C.H.; Schinkel, A.F.L.; Bosch, J.G.; de Jong, N.; van Soest, G.; van der Steen, A.F.W. High-definition imaging of carotid artery wall dynamics. *Ultrasound Med. Biol.* **2014**, *40*, 2392–2403. [CrossRef] [PubMed]

47. Kruizinga, P.; Mastik, F.; Bosch, J.G.; De Jong, N.; Van Der Steen, A.F.W.; Van Soest, G. Measuring submicrometer displacement vectors using high-frame-rate ultrasound imaging. *IEEE Trans. Ultrason. Ferroelectr. Freq. Control* **2015**, *62*, 1733–1744. [CrossRef] [PubMed]

48. Li, F.; He, Q.; Huang, C.; Liu, K.; Shao, J.; Luo, J. High frame rate and high line density ultrasound imaging for local pulse wave velocity estimation using motion matching: A feasibility study on vessel phantoms. *Ultrasonics* **2016**, *67*, 41–54. [CrossRef] [PubMed]

49. Strachinaru, M.; Bosch, J.G.; van Dalen, B.M.; van Gils, L.; van der Steen, A.F.W.; de Jong, N.; Geleijnse, M.L.; Vos, H.J. Cardiac shear wave elastography using a clinical ultrasound system. *Ultrasound Med. Biol.* **2017**, *43*, 1596–1606. [CrossRef] [PubMed]

50. Ishii, T.; Yiu, B.Y.S.; Yu, A.C.H. Vector flow visualization of urinary flow dynamics in a bladder outlet obstruction model. *Ultrasound Med. Biol.* **2017**, *43*, 2601–2610. [CrossRef] [PubMed]

51. Cikes, M.; Tong, L.; Sutherland, G.R.; D'hooge, J. Ultrafast cardiac ultrasound imaging: Technical principles, applications, and clinical benefits. *JACC Cardiovasc. Imaging* **2014**, *7*, 812–823. [CrossRef] [PubMed]

52. Tong, L.; Ramalli, A.; Tortoli, P.; Fradella, G.; Caciolli, S.; Luo, J.; D'hooge, J. Wide-angle tissue doppler imaging at high frame rate using multi-line transmit beamforming: An experimental validation in vivo. *IEEE Trans. Med. Imaging* **2016**, *35*, 521–528. [CrossRef] [PubMed]

53. Faurie, J.; Baudet, M.; Assi, K.C.; Auger, D.; Gilbert, G.; Tournoux, F.; Garcia, D. Intracardiac vortex dynamics by high-frame-rate Doppler vortography—In vivo comparison with vector flow mapping and 4-D flow MRI. *IEEE Trans. Ultrason. Ferroelectr. Freq. Control* **2016**, *64*, 424–432. [CrossRef] [PubMed]

54. Van Cauwenberge, J.; Lovstakken, L.; Fadnes, S.; Rodriguez-Morales, A.; Vierendeels, J.; Segers, P.; Swillens, A. Assessing the performance of ultrafast vector flow imaging in the neonatal heart via multiphysics modeling and in vitro experiments. *IEEE Trans. Ultrason. Ferroelectr. Freq. Control* **2016**, *63*, 1772–1785. [CrossRef] [PubMed]

55. Tremblay-Darveau, C.; Williams, R.; Milot, L.; Bruce, M.; Burns, P.N. Visualizing the tumor microvasculature with a nonlinear plane-wave Doppler imaging scheme based on amplitude modulation. *IEEE Trans. Med. Imaging* **2016**, *35*, 699–709. [CrossRef] [PubMed]

56. Correia, M.; Provost, J.; Tanter, M.; Pernct, M. 4D ultrafast ultrasound flow imaging: In vivo quantification of arterial volumetric flow rate in a single heartbeat. *Phys. Med. Biol.* **2016**, *61*, L48–L61. [CrossRef] [PubMed]

57. Hansen, K.; Udesen, J.; Gran, F.; Jensen, J.; Bachmann Nielsen, M. In-vivo examples of flow patterns with the fast vector velocity ultrasound method. *Ultraschall Med.* **2009**, *30*, 471–477. [CrossRef] [PubMed]

58. Rossini, L.; Martinez-Legazpi, P.; Vu, V.; Fernández-Friera, L.; Pérez del Villar, C.; Rodríguez-López, S.; Benito, Y.; Borja, M.G.; Pastor-Escuredo, D.; Yotti, R.; et al. A clinical method for mapping and quantifying blood stasis in the left ventricle. *J. Biomech.* **2016**, *49*, 2152–2161. [CrossRef] [PubMed]

59. Hansen, K.L.; Møller-Sørensen, H.; Kjaergaard, J.; Jensen, M.B.; Lund, J.T.; Pedersen, M.M.; Lange, T.; Jensen, J.A.; Nielsen, M.B. Analysis of systolic backflow and secondary helical blood flow in the ascending aorta using vector flow imaging. *Ultrasound Med. Biol.* **2016**, *42*, 899–908. [CrossRef] [PubMed]

60. Hansen, K.L.; Udesen, J.; Oddershede, N.; Henze, L.; Thomsen, C.; Jensen, J.A.; Nielsen, M.B. In vivo comparison of three ultrasound vector velocity techniques to MR phase contrast angiography. *Ultrasonics* **2009**, *49*, 659–667. [CrossRef] [PubMed]

61. Swillens, A.; Segers, P.; Torp, H.; Løvstakken, L. Two-dimensional blood velocity estimation with ultrasound: Speckle tracking versus crossed-beam vector doppler based on flow simulations in a carotid bifurcation model. *IEEE Trans. Ultrason. Ferroelectr. Freq. Control* **2010**, *57*, 327–339. [CrossRef] [PubMed]

62. Chee, A.J.Y.; Ho, C.K.; Yiu, B.Y.S.; Yu, A.C.H. Walled Carotid Bifurcation Phantoms for Imaging Investigations of Vessel Wall Motion and Blood Flow Dynamics. *IEEE Trans. Ultrason. Ferroelectr. Freq. Control* **2016**, *63*, 1852–1864. [CrossRef] [PubMed]

63. Hoskins, P.R. Simulation and Validation of Arterial Ultrasound Imaging and Blood Flow. *Ultrasound Med. Biol.* **2008**, *34*, 693–717. [CrossRef] [PubMed]

64. Greyling, A.; van Mil, A.C.C.M.; Zock, P.L.; Green, D.J.; Ghiadoni, L.; Thijssen, D.H. Adherence to guidelines strongly improves reproducibility of brachial artery flow-mediated dilation. *Atherosclerosis* **2016**, *248*, 196–202. [CrossRef] [PubMed]

65. Failla, M.; Grappiolo, A.; Emanuelli, G.; Vitale, G.; Fraschini, N.; Bigoni, M.; Grieco, N.; Denti, M.; Giannattasio, C.; Mancia, G. Sympathetic tone restrains arterial distensibility of healthy and atherosclerotic subjects. *J. Hypertens.* **1999**, *17*, 1117–1123. [CrossRef] [PubMed]

66. Swierblewska, E.; Hering, D.; Kara, T.; Kunicka, K.; Kruszewski, P.; Bieniaszewski, L.; Boutouyrie, P.; Somers, V.K.; Narkiewicz, K. An independent relationship between muscle sympathetic nerve activity and pulse wave velocity in normal humans. *J. Hypertens.* **2010**, *28*, 979–984. [CrossRef] [PubMed]

67. Van Bortel, L.M.; Laurent, S.; Boutouyrie, P.; Chowienczyk, P.; Cruickshank, J.K.; De Backer, T.; Filipovsky, J.; Huybrechts, S.; Mattace-Raso, F.U.S.; Protogerou, A.D.; et al. Expert consensus document on the measurement of aortic stiffness in daily practice using carotid-femoral pulse wave velocity. *J. Hypertens.* **2012**, *30*, 445–448. [CrossRef] [PubMed]

68. Thijssen, D.H.J.; Black, M.A.; Pyke, K.E.; Padilla, J.; Atkinson, G.; Harris, R.A.; Parker, B.; Widlansky, M.E.; Tschakovsky, M.E.; Green, D.J. Assessment of flow-mediated dilation in humans: A methodological and physiological guideline. *Am. J. Physiol. Heart Circ. Physiol.* **2011**, *300*, H2–H12. [CrossRef] [PubMed]

69. Papamichael, C.M.; Aznaouridis, K.A.; Karatzis, E.N.; Karatzi, K.N.; Stamatelopoulos, K.S.; Vamvakou, G.; Lekakis, J.P.; Mavrikakis, M.E. Effect of coffee on endothelial function in healthy subjects: The role of caffeine. *Clin. Sci.* **2005**, *109*, 55–60. [CrossRef] [PubMed]

70. Hijmering, M.L.; de Lange, D.W.; Lorsheyd, A.; Kraaijenhagen, R.J.; van de Wiel, A. Binge drinking causes endothelial dysfuntion, which is not prevented by wine polyphenols: A small trial in healthy volunteers. *Neth. J. Med.* **2007**, *65*, 29–35. [PubMed]

71. Dawson, E.A.; Whyte, G.P.; Black, M.A.; Jones, H.; Hopkins, N.; Oxborough, D.; Gaze, D.; Shave, R.E.; Wilson, M.; George, K.P.; et al. Changes in vascular and cardiac function after prolonged strenuous exercise in humans. *J. Appl. Physiol.* **2008**, *105*, 1562–1568. [CrossRef] [PubMed]

72. Tinken, T.M.; Thijssen, D.H.J.; Hopkins, N.; Black, M.A.; Dawson, E.A.; Minson, C.T.; Newcomer, S.C.; Laughlin, M.H.; Cable, N.T.; Green, D.J. Impact of shear rate modulation on vascular function in humans. *Hypertension* **2009**, *54*, 278–285. [CrossRef] [PubMed]

73. Doonan, R.J.; Hausvater, A.; Scallan, C.; Mikhailidis, D.P.; Pilote, L.; Daskalopoulou, S.S. The effect of smoking on arterial stiffness. *Hypertens. Res.* **2010**, *33*, 398–410. [CrossRef] [PubMed]

74. Ahuja, K.D.; Robertson, I.K.; Ball, M.J. Acute effects of food on postprandial blood pressure and measures of arterial stiffness in healthy humans. *Am. J. Clin. Nutr.* **2009**, *90*, 298–303. [CrossRef] [PubMed]

75. Novak, V.; Novak, P. Influence of respiration on heart rate and blood pressure fluctuations. *J. Appl. Physiol.* **1993**, *74*, 617–626. [CrossRef] [PubMed]

76. Ginghina, C.; Beladan, C.C.; Iancu, M.; Calin, A.; Popescu, B.A. Respiratory maneuvers in echocardiography: A review of clinical applications. *Cardiovasc. Ultrasound* **2009**, *7*, 42. [CrossRef] [PubMed]

77. Cinthio, M.; Ahlgren, Å.R.; Bergkvist, J.; Jansson, T.; Persson, H.W.; Lindström, K. Longitudinal movements and resulting shear strain of the arterial wall. *Am. J. Physiol. Heart Circ. Physiol.* **2006**, *291*, H394–H402. [CrossRef] [PubMed]

78. Schulz, K.F.; Altman, D.G.; Moher, D.; Jüni, P.; Altman, D.; Egger, M.; Chan, A.; Altman, D.; Glasziou, P.; Meats, E.; et al. CONSORT 2010 Statement: Updated guidelines for reporting parallel group randomised trials. *BMC Med.* **2010**, *8*, 18. [CrossRef] [PubMed]

79. Hines, E.A.; Brown, G.E. The cold pressor test for measuring the reactibility of the blood pressure: Data concerning 571 normal and hypertensive subjects. *Am. Heart J.* **1936**, *11*, 1–9. [CrossRef]

80. Saab, P.G.; Llabre, M.M.; Hurwitz, B.E.; Schneiderman, N.; Wohlgemuth, W.; Durel, L.A.; Massie, C.; Nagel, J. The cold pressor test: Vascular and myocardial response patterns and their stability. *Psychophysiology* **1993**, *30*, 366–373. [CrossRef] [PubMed]

81. Kline, K.A.; Saab, P.G.; Llabre, M.M.; Spitzer, S.B.; Evans, J.D.; McDonald, P.A.G.; Schneiderman, N. Hemodynamic response patterns: Responder type differences in reactivity and recovery. *Psychophysiology* **2002**, *39*, 739–746. [CrossRef] [PubMed]

82. Flück, D.; Ainslie, P.N.; Bain, A.R.; Wildfong, K.W.; Morris, L.E.; Fisher, J.P. Extra- and intracranial blood flow regulation during the cold pressor test: Influence of age. *J. Appl. Physiol.* **2017**, *123*, 1071–1080. [CrossRef] [PubMed]

83. Au, J.S.; Bochnak, P.A.; Valentino, S.E.; Cheng, J.L.; Stöhr, E.J. Cardiac and haemodynamic influence on carotid artery longitudinal wall motion. *Exp. Physiol.* **2018**, *103*, 141–152. [CrossRef] [PubMed]

84. Lo, A.; Oehley, M.; Bartlett, A.; Adams, D.; Blyth, P.; Al-Ali, S. Anatomical variations of the common carotid artery bifurcation. *AZN J. Surg.* **2006**, *76*, 970–972. [CrossRef] [PubMed]

85. Schulz, U.G.R.; Rothwell, P.M. Major variation in carotid bifurcation anatomy: A possible risk factor for plaque development? *Stroke* **2001**, *32*, 2522–2529. [CrossRef] [PubMed]

86. Hertzog, M.A. Considerations in determining sample size for pilot studies. *Res. Nurs. Health* **2008**, *31*, 180–191. [CrossRef] [PubMed]

87. Pepe, M.S. Receiver operating characteristic methodology. *J. Am. Stat. Assoc.* **2000**, *95*, 308–311. [CrossRef]

MDPI

St. Alban-Anlage 66

4052 Basel

Switzerland

Tel. +41 61 683 77 34

Fax +41 61 302 89 18

www.mdpi.com

*Applied Sciences* Editorial Office

E-mail: applsci@mdpi.com

www.mdpi.com/journal/applsci

www.ingramcontent.com/pod-product-compliance
Lightning Source LLC
Chambersburg PA
CBHW051856210326
41597CB00033B/5922